Eugene P. Northrop ....nnecticut, in 1908. After gradua........om the local high school he attended Robert College, Istanbul, where his uncle, Professor of Mathematics, converted him to a lifelong interest in the subject. Mr Northrop then took degrees of B.S., M.A., and Ph.D. at Yale, and also taught there.

From 1935 to 1943 he was mathematics master at Hotchkiss School, a private boarding school for boys. In 1943 he joined the College Faculty at the University of Chicago at the time when it was becoming nationally famous for its enlightened and adventurous approach to higher education. There, with a staff of young and energetic colleagues, he developed a radically different course in mathematics which has strongly influenced both college and secondary school teaching in the U.S.A. In 1953 he was named William Rainey Harper Professor.

In 1954–5 he was consultant in Washington to the National Science Foundation, and also to the Fund for the Advancement of Education. Since 1960 he has been the Ford Foundation's resident representative in Turkey, at the same time advising on developments in mathematics and science.

EUGENE P. NORTHROP

RIDDLES
IN MATHEMATICS

—

A BOOK OF PARADOXES

Penguin Books

Penguin Books Ltd, Harmondsworth, Middlesex, England
Penguin Books, 625 Madison Avenue, New York, New York 10022, U.S.A.
Penguin Books Australia Ltd, Ringwood, Victoria, Australia
Penguin Books Canada Ltd, 2801 John Street, Markham, Ontario, Canada L3R 1B4
Penguin Books (N.Z.) Ltd, 182–190 Wairau Road, Auckland 10, New Zealand

—

First published 1944
First published in Great Britain by The English Universities Press 1945
Published in Pelican Books 1960
Reprinted with revisions 1961
Reprinted 1963, 1964, 1967, 1971, 1974, 1975, 1978

—

Copyright 1944 by Eugene P. Northrop
All rights reserved

—

Made and printed in Great Britain
by Cox & Wyman Ltd,
London, Reading and Fakenham
Set in Monotype Times

To
B. U. N.

Contents

are equal. Division by zero in disguise. Peculiar proportions. Contradictions in equations. Plus and minus. Any number is one greater than itself. Inequalities. Any number is any amount greater than itself. An eighth is greater than a fourth. Imaginary numbers. Minus one equal to plus one in two instances.

CONTENTS

Preface

POPULAR interest in mathematics is unquestionably increasing. Perhaps this is because of the fact that mathematics is a tool without which the applied sciences would cease to be sciences. On the other hand, the abstract aspect of mathematics is beginning to attract a large following of people who, weary of the complexities of the human equation in everyday activities, turn in their leisure to the simplicities of the mathematical equation. It is for these people that this book is written. Indeed, only two things are required of the prospective reader – an elementary training in mathematics, and an interest in matters mathematical. These two prerequisites are sufficient for an understanding of the first nine chapters of the book. The tenth – and last – chapter is specifically designed for the reader with more technical equipment.

Of all the problems dealt with in mathematics, paradoxes are among the most appealing and instructive. The appeal of a paradox is difficult to analyse in a word or two, but it probably arises from the fact that a contradiction comes as a complete surprise in what is generally thought of as the only 'exact' science. And a paradox is always instructive, for to unravel the troublesome line of reasoning requires a close scrutiny of the fundamental principles involved. In the light of these arguments it has seemed worth while to bring out a book devoted exclusively to some of the paradoxes which mathematicians, both amateur and professional, have found disconcerting.

The material for this book has been gathered from a wide variety of sources. Some of it has naturally appeared in other popular expositions of mathematics – such works as Ball's *Mathematical Recreations and Essays*, Steinhaus's *Mathematical Snapshots*, and Kasner and Newman's *Mathematics and the Imagination*, to mention only three. If this is a fault, it is not the fault of the author, but of the material he is trying to present. An attempt is made, in the majority of instances, to give references to original sources. This is not always possible, however, particularly when the same problem, in different forms, is to be found in a number of different places.

The author wishes to express his thanks to all who have

contributed to the development of this book. He is particularly indebted to Mr Henry C. Edgar, of the Hotchkiss School, for his painstaking study and criticism of the entire manuscript. Without his help many points, clear enough to the mathematician, would have remained obscure to the general reader. Special thanks are also due to the author's former teacher and colleague, Professor Einar Hille, of Yale University, who read and criticized the manuscript from the point of view of the mathematician.

E. P. NORTHROP

Chicago, Illinois

CHAPTER ONE

What is a Paradox?

Two fathers and two sons leave town. This reduces the population of the town by three. False? No, true – provided the trio consists of father, son, and grandson.

A bookworm starts at the outside of the front cover of volume I of a certain set of books and eats his way to the outside of the back cover of volume III. If each volume is one inch thick, he must travel three inches in all. True? No, false. A moment's study of the accompanying figure shows that he has only to make his way through volume II – a distance of one inch.

FIG. 1

A man says, 'I am lying.' Is his statement true? If so, then he is lying, and his statement is false. Is his statement false? If so, then he is lying, and his statement is true.

The dictionaries define an island as 'a body of land completely surrounded by water' and a lake as 'a body of water completely surrounded by land'. But suppose the northern hemisphere were all land, and the southern hemisphere all water. Would you call the northern hemisphere an island, or would you call the southern hemisphere a lake?

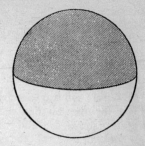

FIG. 2. If the northern hemisphere were all land and the southern
hemisphere all water

It is of such brain-twisters as these that this book is composed.
There are paradoxes for everyone – from the person who left
mathematics behind in school (or who was left behind in school
by mathematics) to the professional mathematician, who is still
bothered by such a problem as that of the liar.

We shall use the word 'paradox', by the way, in the sense in
which it is used in these examples. That is to say, a paradox is anything which offhand appears to be false, but is actually true; or
which appears to be true, but is actually false; or which is simply
self-contradictory. From time to time it may appear that we are
straying from this meaning. But be patient – what seems crystal-
clear to you may leave the next person completely confused.

*

If you are among those who at this point are saying, 'But we
thought this book had to do with *mathematical* paradoxes – how
about it?' then stay with us for a moment. If you are not interested in the answer to this question, you may as well skip to the
next chapter.

A closer look at the difficulties encountered in our first examples
will show that they are simply cases of very real difficulties encountered not only by the student of mathematics, but by the
mature mathematician as well.

In the problem concerning fathers and sons, we find ourselves
searching here and there for some instance in which the conditions
of the problem will be fulfilled. It seems at first as though such an
instance cannot possibly exist – common sense and intuition are

all against it. But suddenly, there it is – as simple a solution as can be. This sort of thing happens time and again in mathematical research. The mathematician, working on the development of some theory or other, is suddenly confronted with a set of conditions which appear to be highly improbable. He begins looking for an example to fit the conditions, and it may be days, or weeks, or even longer, before he finds one. Frequently the solution of his difficulty is as simple as was ours – the kind of thing that makes him wonder why he hadn't thought of it before.

The problem of the bookworm's journey is a nice example of the way in which reason can be led astray by hasty judgement. The false conclusion is reached through failure to investigate carefully all aspects of the problem. There are many specimens of this sort – much more subtle ones, to be sure – in the literature of mathematics. A number of them enjoyed careers lasting many years before some doubting mathematician finally succeeded in discovering the trouble.

The case of the self-contradicting liar is but one of a whole string of logical paradoxes of considerable importance. Invented by the early Greek philosophers, who used them chiefly to confuse their opponents in debate, they have in more recent times served to bring about revolutionary changes in ideas concerning the nature and foundations of mathematics. In a later chapter we shall have more to say about problems of this kind.

The island-and-lake problem, which had to do with definitions and reasoning from definitions, is really typical of the development of any mathematical theory. The mathematician first defines the objects with which he is going to work – numbers, or points, or lines, or even just 'elements' of an unspecified nature. He then lays down certain laws – 'axioms', he calls them, or 'postulates' – which are to govern the behaviour of the objects he has defined. On this foundation he builds, through a series of logical arguments, a whole structure of mathematical propositions, each one resting on the conclusions established before it. He is not interested, by the way, in the *truth* of his definitions or axioms, but asks only that they be *consistent*, that is, that they lead to no real contradiction in the propositions (such, for example, as the contradiction in the problem of the liar). Bertrand Russell, in his *Mysticism and Logic*, has put what we are trying to say in the following words:

Pure mathematics consists entirely of assertions to the effect that if such and such a proposition is true of anything, then such and such another proposition is true of that thing. It is essential not to discuss whether the first proposition is really true, and not to mention what the anything is of which it is supposed to be true. . . . Thus mathematics may be defined as the subject in which we never know what we are talking about, nor whether what we are saying is true.

How is that, by the way, for a paradox?

CHAPTER TWO

Paradoxes for Everyone

MANY of the anecdotes and problems of this chapter are fairly well known. All of them have probably appeared in print in some form or other and at some time or other, and a few are so common that they can be found in almost any book on mathematical puzzles and games. It is next to useless to try to trace them to their original sources – most of them, like Topsy, 'just growed'.[1]*

*

We shall begin with a couple of lessons in geography. The first concerns a man who, you will say, must have been a crank. He designed a square house with windows on all four sides, each window having a view to the south. No bay windows (which would take care of three sides) or anything of that sort. Now how on earth can this be done? *Where* on earth would be more to the point, for there is indeed only one place where such a house can be built. Does that give it away? You've got it – it's the North Pole, of course, from which any direction is south.

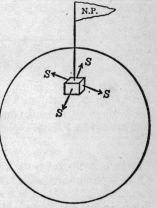

FIG. 3. Any direction from the North Pole is south

Without the foregoing discussion, the following problem strikes most people as quite paradoxical. A certain sportsman, experienced in shooting small game, was out on his first bear hunt. Suddenly he spotted a huge bear about a hundred yards due *east* of him. Seized with panic, the hunter ran – not directly away

---

* See page 227. Notes and references for all chapters will be found near the end of the book. They are inserted for the convenience of all who are interested in them and can be ignored safely by all who are not.

from the bear, but, in his confusion, due *north*. Having covered about a hundred yards, he regained his presence of mind, stopped, turned, and killed the bear – who had not moved from his original position – by shooting due *south*. Have you all the data clearly in mind? Very well, then; what colour was the bear?

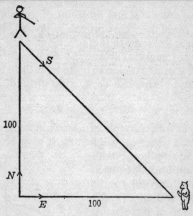

FIG. 4. Details of the bear hunt

The same problem can be put in another, although perhaps less startling form. Where can a man set out from his house, walk five miles due south, five miles due west, and five miles due north and find himself back home?

*

Charles L. Dodgson, better known to the general public as Lewis Carroll, the author of *Alice in Wonderland*, is recognized by mathematicians and logicians as one of their own number. We are indebted to him for the following paradox,[2] as well as for several others which appear in later parts of the book.

We can agree, can we not, that the better of two clocks is the one that more often shows the correct time? Now suppose we are offered our choice of two clocks, one of which loses a minute a day, while the other does not run at all. Which one shall we accept? Common sense tells us to take the one that loses a minute a day, but if we are to stick to our agreement, we shall have to take the one that doesn't run at all. Why? Well, the clock that loses a minute a day, once properly set, will have to lose 12 hours, or 720 minutes, before it is right again. And if it loses only a minute a day, it will take 720 days to lose 720 minutes. In other words, it is correct only once about every two years. But the clock that doesn't run at all is correct twice a day!

*

Apparently impossible results are frequently obtained through either too little attention to relevant details or too much attention to irrelevant ones. Let's look at a few problems of this kind. We shan't bother, by the way, to discuss their solutions here.

FIG. 5. Equal in value or not?

A scatter-brained young lady once went into a jeweller's shop, picked out a ring worth £1, paid for it, and left. She appeared at the shop the next day and asked if she might exchange it for another. This time she picked out one worth £2, thanked the jeweller sweetly, and started to leave. He naturally demanded an additional £1. The young lady indignantly pointed out that she had paid him £1 the day before, that she had just returned to him a £1 ring, and that she therefore owed him nothing. Whereupon she stalked out of the shop and left the jeweller wildly counting on his fingers.

Then there is the story of the young man who once found himself applying for a job. He told the manager that he thought he was worth £300 a year to him. The manager apparently thought otherwise. 'Look here,' he said, 'there are 365 days in the year. You sleep 8 hours a day, or a total of 122 days. That leaves 243.

You rest 8 hours a day, or a total of 122 days. That leaves 121. You do no work for 52 Sundays. That leaves 69 days. You have half a day off on 52 Saturdays – a total of 26 days. That leaves 43. You have an hour off for lunch each day – a total of 15 days. That leaves 28. You have a fortnight's holiday. That leaves 14 days. And then come Easter Monday, Whit-Monday, August Bank Holiday, and Christmas. Do you think you're worth £300 to me for 10 working days?'

A group of seven weary men once arrived at a small hotel and asked for accommodation for the night, specifying that they wanted separate rooms. The manager admitted that he had only six rooms left, but thought he might be able to put up his guests as they desired. He took the first man to the first room and asked one of the other men to stay there for a few minutes. He then took the third man to the second room, the fourth man to the third room, the fifth man to the fourth room, and the sixth man to the fifth room. Then he returned to the first room, got the seventh man, and showed him to the sixth room. Everyone was thus nicely taken care of. Or was he?

Here is another problem of this type, not quite so simple. Three men had dinner at a hotel, received a bill for 30 shillings, and each handed a 10-shilling note to the waiter. He took the money to the office, where he was told that there had been a mistake – the bill should have been for 25 shillings, not 30; so he was sent back with 5 shillings. On the way back it occurred to him that 5 shillings was going to be difficult to divide between three men, that the men did not know the actual amount of the bill anyway, and that they would be glad of any return on the money. So he kept 2 shillings and returned one to each of the three men. Now each of the men paid 9 shillings. Three times 9 is 27. The waiter had 2 shillings in his pocket. 27 plus 2 is 29, and the men originally handed over 30 shillings. Where *is* that other shilling?

While we are on the subject of money, there is that very puzzling story having to do with foreign exchange. The governments of two neighbouring countries – let's call them Northia and Southia – had an agreement whereby a Northian dollar was worth a dollar in Southia, and vice versa. But one day the government of Northia decreed that thereafter a Southian dollar was to be worth but ninety cents in Northia. The next day the Southian government, not to be outdone, decreed that thereafter a Northian dollar was to be worth

18

but ninety cents in Southia. Now a bright young man lived in a town which straddled the border between the two countries. He went into a store on the Northian side, bought a ten-cent razor, and paid for it with a Northian dollar. He was given a Southian dollar, worth ninety cents there, in change. He then crossed the street, went into a Southian store, bought a ten-cent package of blades, and paid for it with the Southian dollar. There he was given a Northian dollar in change. When the young man returned home, he had his original dollar *and* his purchases. And each of the tradesmen had ten cents in his cash-drawer. Who, then, paid for the razor and blades?

*

One of the oldest paradoxes is that of the wealthy Arab who at death left his stable of seventeen beautiful horses to his three sons. He specified that the eldest was to have one half of the horses, the next one-third, and the youngest one-ninth. The three young heirs were in despair, for they obviously could not divide seventeen horses this way without calling in the butcher. They finally sought the advice of an old and wise friend, who promised to help them. He arrived at the stable the next day, leading one of his own horses. This he added to the seventeen and directed the brothers to make their choices. The eldest took one half of the eighteen, or nine; the next, one-third of the eighteen or six; and the youngest, one-ninth of the eighteen, or two. When all seventeen of the original horses had been chosen,

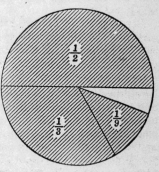

FIG. 6. The fractions $\frac{1}{2}$, $\frac{1}{3}$, and $\frac{1}{9}$ do not total 1

the old man took his own horse and departed. The catch? It's in the father's stipulations. Either he was a poor arithmetician or he wanted to give his sons something to think about. At any rate, the fractions one-half, one-third, and one-ninth do not add up to one – as they should if nothing is to be left over – but to seventeen-eighteenths.

*

A large business firm was once planning to open a new branch in a certain city, and advertised positions for three clerks. Out of a number of applicants the personnel manager selected three promising young men and addressed them in the following way:

'Your salaries are to begin at the rate of £200 per year, to be paid every half-year. If your work is satisfactory, and we keep you, your salaries will be raised. Which would you prefer, a rise of £30 per year or a rise of £10 every half-year?' The first two of the three applicants eagerly accepted the first alternative, but the third young man, after a moment's reflection, took the second. He was promptly put in charge of the other two. Why? Was it because the personnel manager liked his modesty and apparent willingness to save the company money? Not at all. As befitting his position, he actually received more salary than his companions. They had jumped to the conclusion that a rise of £10 every half-year was equivalent to a rise of £20 per year, but he had taken all the conditions of the problem into consideration. He had lined up the two possibilities and had looked at the yearly salaries in this way:

	£30 rise yearly	£10 rise half-yearly
1st year	£100+£100 = £200	£100+£110 = £210
2nd year	115+ 115 = 230	120+ 130 = 250
3rd year	130+ 130 = 260	140+ 150 = 290
4th year	145+ 145 = 290	160+ 170 = 330

It was then immediately apparent to him that his salary in succeeding years would exceed theirs by £10, 20, 30, 40, and so on, his rise each year exceeding theirs by £10. It was his alertness of mind, and not his modesty, that impressed his new employer.

*

Most people are easily confused by problems involving average rates of speed. Try this one on your friends.

A man drove his car 1 mile to the top of a mountain at the rate of 15 miles per hour. How fast must he drive 1 mile down the other side in order to average 30 miles per hour for the whole trip of 2 miles?

First let us look at it in this way: he would average 30 miles per hour for the whole trip if he drove the second mile at the rate of 45 miles per hour, for the average of 15 and 45 is (15+45)/2, or 30.

But now suppose we look at it in another way. Using our old

friend, the relation 'distance = rate × time', we note that the time required to drive 2 miles at the average rate of 30 miles per hour is $\frac{2}{30}$ of an hour, or 4 minutes. Furthermore, the time required to drive 1 mile at the rate of 15 miles per hour is $\frac{1}{15}$ of an hour, or again 4 minutes. In other words, our traveller must cover that second mile in 0 seconds flat!

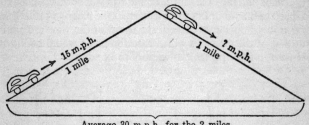

Average 30 m.p.h. for the 2 miles

FIG. 7

Which of these results are we to accept? The second is the correct one, and shows that considerable care must be used in averaging rates. The average rate for any trip is always found by dividing the *total distance* by the *total time*. In our first analysis, if the man drives one mile at 15 miles per hour and a second mile at 45 miles per hour, the times for those two miles are $\frac{1}{15}$ and $\frac{1}{45}$ of an hour respectively, or $\frac{4}{45}$ of an hour in all. His average rate is thus $2/\frac{4}{45}$ or 22·5 miles per hour. This discussion should furnish a practical tip to those drivers who allow just so much time to get somewhere. They cannot average 50 miles per hour, for example, by going a certain number of *miles* at 40 miles per hour and the same number of *miles* at 60 miles per hour. On the other hand, they *can* average 50 by going 40 and 60 for the same number of *hours*. For if they maintain these respective rates for one hour each, they will have gone 100 miles in 2 hours.

With the help of the above discussion you ought to be able to pick out the flaws in the following two arguments. If not, you will find their solutions in the Appendix towards the end of the book.

Paradox 1. A plane makes a trip from London to Liverpool and back to London. Call the distance between the two cities 200 miles and the speed of the plane 100 miles per hour. Then the time required for the round trip, ignoring stops, is 4 hours. Now suppose there is a strong wind which blows throughout the entire trip with the same speed and in the same direction – from London directly towards Liverpool, say. Then the tail wind on the way south will speed up the plane to the same extent that the head wind will retard it on the way north. In other words, both the average speed of the plane and the time for the round trip will be independent of the speed of the wind. But this means that the plane can still make the trip in 4 hours even though the speed of the wind is greater than that of the plane, in which case the plane would be blown *backwards* on the trip from Liverpool to London!

30 at 2 a 1d.

30 at 3 a 1d.

60 at 5 for 2d.

FIG. 8

Paradox 2. Each of two apple women had 30 apples for sale. The first sold hers at the rate of 2 a penny, the second at the rate of 3 a penny. When the apples were sold their respective receipts were 15 pence and 10 pence, or 25 pence in all. Next time the women decided to do business together, so they pooled their 60 apples and sold them at the rate of 5 for twopence (2 a penny plus 3 a penny). Upon counting their joint receipts they were dismayed to find that they had only 24 pence. They searched all about them for that other penny and wound up by bitterly accusing each other of having taken it. Where was it?

*

There are many problems in which the obvious solution is never the correct one. That is to say, what offhand appears to be true is false. The following four deserve mention, although they are pretty well known. As in the case of the last two paradoxes, their correct solutions are given in the Appendix.

Paradox 3. A clock strikes six in 5 seconds. How long does it take to strike twelve? No! The answer is *not* 10 seconds.

Paradox 4. A bottle and its cork cost together 1*s* 1*d*. The bottle costs a shilling more than the cork. How much does the bottle cost? No! The answer is *not* 1*s*.

Paradox 5. A frog is at the bottom of a 30-foot well. Each hour he climbs 3 feet and slips back 2. How many hours does it take him to get out? No! The answer is *not* 30 hours.

Paradox 6. An express leaves London for Brighton at the same time as a slow train leaves Brighton for London. The express travels at the rate of 50 miles per hour, the slow train at the rate of 30 miles per hour. Which is farther from London when they meet? No! The answer is *not* the express.

*

Two problems, similar to the last four, had better be taken up in detail here.

A farmer's wife once drove to town to sell a basket of eggs. To her first customer she sold half her eggs and half an egg. To the second customer she sold half of what she had left and half an egg. And to the third customer she sold half of what she *then* had left and half an egg. Three eggs remained. How many did she start out with? Now the only thing which makes this problem paradoxical is this additional condition: *she didn't break any eggs.* It takes only a moment's reflection, though, to see that this condition will be fulfilled if she starts with an *odd* number of eggs. The answer is 31.

And now for our second problem. Let's suppose that we have in one glass a certain quantity of water and in another glass an equal quantity of milk. We shall assume, by the way, that this is good, old-fashioned, unwatered milk. We take a teaspoonful of milk from the first glass, put it in the second, and stir. We then take a teaspoonful of the mixture from the second glass and put it back in the first glass. Now is there more water in the milk than milk in the water, or more milk in the water than water in the milk?

The people to whom this problem is proposed generally split up into two groups. On the one hand are those who support the first suggestion; on the other hand, those who support the second. Both are wrong. Why? Well, suppose for simplicity, that we start with 4 teaspoonfuls each of milk and water. If we put one teaspoonful of milk in the water, the resulting five teaspoonfuls of mixture

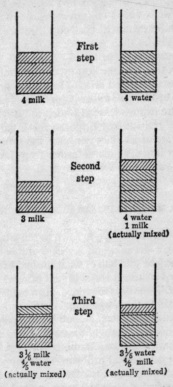

FIG. 9. Details of the milk and water problem

is $\frac{1}{5}$ milk and $\frac{4}{5}$ water. When we transfer one teaspoon of the mixture to the glass of milk, we are returning $\frac{1}{5}$ of a teaspoonful of milk – thus leaving $\frac{4}{5}$ of a teaspoonful of milk in the water – and are adding $\frac{4}{5}$ of a teaspoonful of water to the milk. Thus there are equal quantities – $\frac{4}{5}$ of a teaspoonful – of milk in water and water in milk. Incidentally, it makes no difference whether or not we stir the mixture! Can you see why?

*

We conclude this chapter with a few puzzles involving family relationships.[3] Such puzzles are not, strictly speaking, a part of mathematics, yet the type of reasoning required for their solution is closely akin to the type of reasoning sometimes used by the mathematician.

Take this situation, for example. A big Indian and a little Indian are sitting on a fence. The little Indian is the son of the big Indian, but the big Indian is not the father of the little Indian. What relationship exists between the two? The answer is 'mother and son', but most people fail to get it the first time they hear the story. Their failure can perhaps be traced to the fact that they learn, as children, about division of labour among Indians, and just naturally assume that the squaw doesn't have time to sit on fences, whereas the brave has all the time in the world. The solution of this puzzle, then, requires the ability to dismiss fixed ideas and to look for new ones – a trait which is sometimes of great value to the mathematician.

> Brothers and sisters have I none,
> But that man's father is my father's son

is a fairly well-known riddle and presents no great difficulties. If the speaker is, as he says, an only child, then 'my father's son' is the speaker himself. And if 'that man's father' is 'my father's son', then 'that man's father' is the speaker. Therefore 'that man' is the son of the speaker. All of which not only *sounds* like a demonstration in geometry, but actually *is* like one.

Then there is the complicated family gathering consisting of one grandfather, one grandmother, two fathers, two mothers, four children, three grandchildren, one brother, two sisters, two sons, two daughters, one father-in-law, one mother-in-law, and one daughter-in-law. Let's count them up. Twenty-three people, you say? No, only seven. There were two girls and a boy, their father and mother, and their father's father and mother. A detailed explanation here is literally too much for words. The most satisfactory thing to do is to sit down, write out a list of the seven people involved, and check off the twenty-three relationships.

Surely you have heard of the man who once married his widow's sister. 'Now that', you will reply, 'is utterly impossible. After all, a man's widow does not exist until the man himself *ceases* to exist.' Well, it all happened this way. When the man – let's call him John – was young, he married a girl named Anne. A few years later Anne died. But Anne had a sister, Betty, and John took her for his second wife. Then John died, making Betty his widow. Had not John once married Anne, his widow's sister? Sorry – the catch there was a grammatical one!

In these days of relatively frequent divorces and remarriages it is quite possible for two men, totally unrelated, to have the same sister. A diagram will be of help here.

Mrs *A*——Mr *A*	Mr *A*——Mrs *B*	Mrs *B*——Mr *C*
Son (*AA*)	Daughter (*AB*)	Son (*BC*)

As is indicated, Mr and Mrs *A* had a son, *AA*. Mr and Mrs *A* were then divorced, and Mr *A* proceeded to marry Mrs *B*. These two had a daughter, *AB*. Mr *A* was apparently a difficult man to get along with, for it was not many years before his second wife divorced him and married Mr *C*. A son, *BC*, was born of this last marriage. And now for the dénouement. The two sons, *AA* and *BC*, have no common blood in their veins. They are therefore totally unrelated. Yet each of them is the brother of the daughter, *AB*, for *AA* and *AB* had the same father, Mr *A*, and *AB* and *BC* had the same mother, Mrs *B*.[4]

Everyone has heard of the fallibility of lawmakers and of the laws they make. There is, for example, the choice gem said to have been produced by those who are responsible for railway traffic in one of the south-western states of the U.S.A. It ran something like this: 'If two trains, travelling in opposite directions along the same single track, shall meet one another, neither shall proceed until the other has withdrawn.'

But of all the laws that can lead to extraordinary situations, one of the best – or worst – comes from England. There, between 1907 and 1921, it was possible for a boy to be the legitimate son of his father and, at the same time, the illegitimate son of his mother. For during that period of fourteen years it was legal for a man to marry the sister of his deceased wife, while it was not legal for a

woman to marry the brother of her deceased husband.[5] And here
is what might have happened:

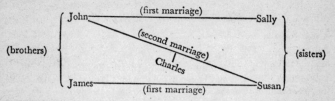

The brothers John and James of our diagram took as their brides
the sisters Sally and Susan. That is to say, John married Sally and
James married Susan. A few years later both James and Sally died,
and John and Susan, after a decent period, were married. Then
John was legally married to Susan, his former sister-in-law, but
Susan was not legally married to John, her former brother-in-law.
Consequently Charles, who was born of this union, was the legi-
timate son of his father and the illegitimate son of his mother.

As a final complication we offer the strange case of two men each
of whom was at the same time both nephew and uncle of the other.
Impossible? No, though perhaps improbable. Here is one solution:

Mr Allen———Mrs Allen———Dick

Tom   Harry

Mr Black———Mrs Black———Tom

Dick   George

In our diagram are Mr and Mrs Allen, who had a son Tom, and
Mr and Mrs Black, who had a son Dick. Mr Allen and Mr Black
both died. And Tom and Dick, after they were grown men, each
married the other's mother. Dick and Mrs Allen then had a son
Harry, and Tom and Mrs Black a son George. Now consider the
relationship between Harry and George. Since Harry is the brother
of Tom, George's father, Harry must be George's uncle. On the
other hand George is the brother of Harry's father, Dick, so Harry
must be George's nephew. In exactly the same way George is
Harry's uncle and nephew.

But this way madness lies.

CHAPTER THREE

Paradoxes in Arithmetic

(NOTE: Throughout this chapter the *billion* used is the Continental or American billion, or one thousand million; *not* the English billion, which is one million million. It may be remarked that the Continental usage is much more convenient than the English, and has in fact recently been adopted by the *Economist*.)

ARITHMETIC is a storehouse of almost unbelievable results. In a single chapter it is possible to discuss only a few of the surprises to be found in this subject, and for the most part we shall confine our attention to some of the remarkable properties of the number 2. There is not much here that the mathematician will find startling – the results to be discussed are paradoxical to the non-mathematician in that he would probably pronounce them false, or at least highly improbable, if he were asked to give his snap judgement on them.

SOME LARGE NUMBERS

In these days of billion-pound government loans and appropriations, most of us have lost our respect for large numbers and are no longer able to appreciate their actual magnitude.

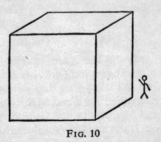

FIG. 10

Just how big *is* a billion, anyway? Well, let's think for the moment of tiny cubical blocks a quarter of an inch each way. A billion such blocks would almost fill a room 21 feet long, 21 feet high, and 21 feet wide.

If spread out in a single layer, they would completely cover 3 full-size football grounds and nearly two-thirds of a fourth.

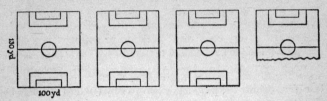

FIG. 11

And if arranged in a straight line, they would reach almost 4000 miles – more than the distance between London and Chicago.

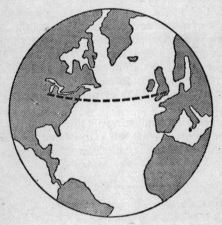

FIG. 12

Or let's think of a billion in connection with time. A billion seconds ago all people now thirty-one years old were not yet born. In 1903 only a billion minutes had elapsed since the birth of Christ. And a billion days ago man was just about to put in an appearance on this earth.

Finally, if your peace of mind is still undisturbed at the thought of the present public debt, consider the fact that in order to pay off a hundred billion pounds at the rate of one pound a second, twenty-four hours a day, seven days a week, and fifty-two weeks a year, it would take about 3,180 years to complete the task!

*

Physicists, chemists, astronomers, and others who deal with large numbers use a very convenient notation in writing them. Note first that a billion is the product of nine 10's. That is to say:

$$1,000,000,000 = 10 \times 10 \times 10 \times 10 \times 10 \times 10 \times 10 \times 10 \times 10$$

Now if we denote the product of two 10's by 10^2, of three 10's by 10^3, of four 10's by 10^4, and so on, then a billion, being the product of nine 10's, can be written as 10^9. Again, four billion can be written as 4×10^9, or 4 with the decimal point moved nine places to the right; 34,870,000,000 as $3 \cdot 487 \times 10^{10}$, or $3 \cdot 487$ with the decimal point moved ten places to the right; and so on.

If we do not wish to be too exact, but merely want some idea of the magnitude of a number, we can say, since $3 \cdot 487$ is nearer to 3 than to 4, that 34,870,000,000 is 'about' 3×10^{10}. If we are after an even rougher approximation, we can say that 3×10^{10} is, 'to the nearest power of 10', 10^{10}. In other words, 3×10^{10} is nearer to 1×10^{10}, or 10^{10}, than to 10×10^{10} or 10^{11}.

What of a number like 4^{3^2}? This is to be interpreted as $4^{(3^2)} = 4^9$, and *not* as $(4^3)^2 = 64^2$.

We can improve our familiarity with this notation by discussing the following problem. What is the largest number which can be written with three 2's? Some possibilities which immediately occur to use are

$$222, \quad 22^2, \quad 2^{22}, \quad \text{and } 2^{2^2}.$$

The smallest of these is $2^{2^2} = 2^4 = 16$. Then come 222, and $22^2 = 484$. The largest is $2^{22} = 4,194,304$, or about 4×10^6.

What if we use four 2's instead of three? The possibilities, arranged in order of increasing magnitude, are now

$$2222, \quad 222^2, \quad 22^{2^2}, \quad 22^{22}, \quad 2^{222}, \quad 2^{22^2}, \quad \text{and } 2^{2^{22}}.$$

The respective values of the first six of these, to the nearest power of 10, are

$$10^3, \quad 10^4, \quad 10^5, \quad 10^{29}, \quad 10^{67}, \quad \text{and } 10^{145}.$$

But the last representation yields $2^{4\,194,304}$, or about $10^{1\,260,000}$. This puts a mere billion to shame, being a billion multiplied by itself some 140,000 times! Did it ever occur to you that four simple 2's could ever amount to that much?

*

Figure 13 is an attempt to illustrate the fact that each person now living had 2 parents, 4 grandparents, 8 great-grandparents, and so on. That is to say, one generation ago he had 2 ancestors. Two

FIG. 13. A family tree in reverse

generations ago he had 4, or 2×2, or 2^2 ancestors. Three generations ago he had 8, or $2 \times 2 \times 2$, or 2^3 ancestors. Four generations ago he had 16, or $2 \times 2 \times 2 \times 2$, or 2^4 ancestors. And so on. In general, n generations ago he had $2 \times 2 \times 2 \times \ldots \times 2$ (the product of n twos), or 2^n ancestors. Now suppose we assume there are 30 years to a generation. Then only 600 years ago – 20 generations back, that is – each one of us had 2^{20}, or 1,040,400 ancestors!

Someone once used this argument to 'prove' that six hundred years ago there were over a million times as many people on this earth as there are today. It doesn't take a census expert to figure out his error. Can you find it?

*

31

The chain letter is an old evil which turns up in some form or other every few years. Consider the simple case in which a person sends a certain letter to two friends, requesting each of them to copy the letter and send it to two of their friends, and so on. Then the first

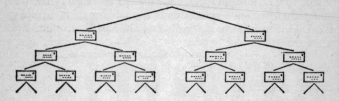

Fig. 14. The chain letter

set consists of two, or 2^1 letters, the second set of four, or 2^2 letters, the third set of eight, or 2^3 letters, and so on. Now how many sets of letters would have to be sent in order that every one of the two billion men, women, and children in the world – literate or illiterate – receive one and only one letter? It is not difficult to show that *it would take no more than thirty sets*! The thirtieth set alone would consist of $2^{30} = 1,073,741,824$ letters.

The thirtieth power of 2 turns up again in the thrifty savings scheme whereby we put away one penny on the first day of the month, two (2^1) pence on the second day, four (2^2) pence on the third day, eight (2^3) pence on the fourth day, and so on – each day doubling the amount of the previous day. Noting that here the power of 2 in each case is one less than the number of the day, it is readily seen that on the thirty-first of the month we should have to put away 2^{30}, or over a billion, pence – more, that is, than four million pounds. The total amount saved would be about twice as much.

*

You should by now be well prepared for the next problem – a good one to try on your friends. Suppose we have a large sheet of very thin rice paper one-thousandth of an inch thick, or a thousand sheets to the inch. We tear the paper in half and put the two pieces together, one on top of the other. We tear them in half and put the four pieces together in a pile, tear *them* in half and put the eight pieces together in a pile, and so on. If we tear and put together fifty times, how high will the final stack of paper be? The usual responses are amusing. Some people suggest a foot, others go as high as several feet, and a few of the bolder ones throw caution to the winds and risk their reputation for sanity on a mile. All of them refuse to believe the correct answer, which is well over *seventeen million miles*!

If you are among the unbelievers, you can work the problem out very simply as follows. As was indicated above, the first tear results in two, or 2^1, pieces of paper; the second tear in four, or 2^2, pieces; the third tear in eight, or 2^3, pieces; and so on. It is evident at once that after the fiftieth tear the stack will consist of 2^{50} sheets of paper. Now 2^{50} is about 1,126,000,000,000,000. And since there are a thousand sheets to the inch, the stack will be 1,126,000,000,000 inches high. To get the height of the stack in feet, divide this number by 12. And to get it in miles, divide the resulting number by 5280. The final result, as we have said, is well over 17,000,000.

*

There are a number of ancient puzzle toys which are to be found even today in many toy-shops. Among them is what is generally known as the 'Tower of Hanoi'. It consists of a horizontal board with three vertical pegs, as shown in Figure 15. On one of the pegs is arranged a series of discs of different sizes, the largest at the

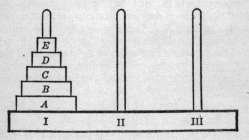

FIG. 15. The Tower of Hanoi

bottom, the next largest on top of that, and so on, up to the smallest at the top of the peg. The problem is to transfer all the discs from the first peg to one of the others – say the third – in such a way that the final arrangement is the same as the original one. *But only one disc is to be moved at a time, and no disc shall ever rest on one smaller than itself.*

For example, suppose the pegs are numbered I, II, and III, and the discs lettered $A, B, C, D, \ldots$ as in the figure. If there are only 2 discs, A and B, then B can be shifted to II, A to III, and B to III. Thus 2 discs require 3, or $2^2 - 1$ transfers. If there are 3 discs, A, B, and C, proceed as follows: C to III, B to II, C to II, A to III, C to I, B to III, and C to III. Thus 3 discs require 7, or $2^3 - 1$ transfers. In general, it can be shown that if there are n discs, a minimum of $2^n - 1$ transfers is required. The game can, of course, be played with discs of cardboard and imaginary pegs. Try it with 5 discs, which require $2^5 - 1 = 31$ moves, and, as you become more proficient, with an even greater number of discs. Here's a helpful hint, by the way. If the number of discs is *even*, move the first disc to the peg numbered II; if it is *odd*, to III.

The origin of the game is described by one author in the following way:[1]

In the great temple at Benares, beneath the dome which marks the centre of the world, rests a brass plate in which are fixed three diamond needles, each a cubit high and as thick as the body of a bee. On one of these needles, at the creation, God placed sixty-four discs of pure gold, the largest disc resting on the brass plate, and the others getting smaller and smaller up to the top one. This is the Tower of Bramah. Day and night unceasingly the priests transfer the discs from one diamond needle to another according to the fixed and immutable laws of Bramah, which require that the priest on duty must not move more than one disc at a time and that he must place this disc on a needle so that there is no smaller disc beneath it. When the sixty-four discs shall have been thus transferred from the needle on which at the creation God placed them to one of the other needles, tower, temple, and Brahmins alike will crumble into dust, and with a thunder-clap the world will vanish.

In this case the number of transfers required is $2^{64} - 1$. If we assume that the priests worked on a 24-hour schedule, transferring discs at the rate of one a second and never making a mistake, it would take them about $5 \cdot 82 \times 10^{11}$ years, or nearly *six billion centuries*, to complete the task. This world's end prophecy is one of the most optimistic on record!

*

The number $2^{64} - 1$ is connected also with the origin of chess. Legend has it that an ancient Shah of Persia was so impressed with the game that he ordered its inventor to ask whatever reward he desired. The inventor – probably a clever arithmetician – asked that he might have one grain of wheat for the first square of a

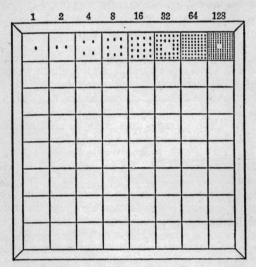

FIG. 16. The chessboard and the grains of wheat

chessboard, two grains for the second square, four grains for the third square, eight grains for the fourth square, and so on, until all the squares of the board were accounted for. Now he was asking for

$$1+2+2^2+2^3+ \ldots +2^{63} = 2^{64}-1$$

grains of wheat. The Shah thought this a poor reward until his advisers worked out the problem for him. They found that $2^{64}-1$ is about $1 \cdot 84 \times 10^{19}$. If it is assumed that there are 9000 grains of wheat to the pint, this figure amounts to some 3×10^{13} bushels, which is several *thousand* times the world's annual crop of wheat even today!

If a second chessboard is placed next to the first, and if the scheme of doubling the number of grains for each successive square is continued, then the pile corresponding to the last square of the second board contains 2^{127} grains. If one grain is removed from this pile, there remain

$$2^{127}-1=170,141,183,460,469,231,731,687,303,715,884,105,727.$$

This number is the *largest known prime number*.[2]

SOME NUMBER THEORY

The last problem brings us to some extraordinary properties of the number 2 which are rather different from those we have been considering. The investigation of prime numbers by both amateur and professional mathematicians has resulted in a wealth of new material for research. We recall that a *prime number* is defined as a number which is exactly divisible by no numbers other than itself and 1. The first twelve prime numbers, for example, are 1, 2, 3, 5, 7, 11, 13, 17, 19, 23, 29, and 31. The number 4 is not prime, for it is divisible by 2. Again, 6 is not prime, for it is divisible by both 2 and 3.

Euclid, the great systematizer of geometry, proved about 300 B.C. that the number of prime numbers is infinite. For many centuries attempts have been made to devise some sort of formula which will generate prime numbers only. For example, the man who first ran across the formula n^2+n+41 must have thought he had something, for this expression yields a prime number when n is any whole number from 1 to 39 inclusive. Thus if n is 1, the formula gives 43; if n is 2, 47; if n is 3, 53; if n is 4, 61; and so on. But if n is 40, the formula gives 1681, which is divisible by 41, being $(41)^2$. This example illustrates the uselessness, in mathematics, of attempting to deduce a general conclusion from a few specific cases.

In 1640 the French mathematician Fermat believed that he had found a formula generating prime numbers only. His suggestion was $2^{2n}+1$, where n is a whole number. The first five 'Fermat numbers', as they are called, are

$$2^{2^0}+1 = 2^1 \ +1 = 3,$$
$$2^{2^1}+1 = 2^2 \ +1 = 5,$$
$$2^{2^2}+1 = 2^4 \ +1 = 17,$$
$$2^{2^3}+1 = 2^8 \ +1 = 257,$$
$$2^{2^4}+1 = 2^{16}+1 = 65,537.$$

(In connection with the first of these numbers, we must recall from algebra that any number raised to the 0th power is 1. See Appendix, page 214.) These are indeed all prime, yet Fermat later began to doubt the truth of his generalization to the effect that his formula will *always* yield a prime number. It was not until a hundred years later that Euler, a Swiss mathematician, found that the sixth Fermat number, $2^{25}+1 = 4,294,967,297$, is the product of 641 and 6,700,417, and so is divisible by either. It has since then been verified that there are other Fermat numbers which are not prime. On the other hand, no one knows as yet whether Fermat's formula gives *any* primes other than the first five which we have seen.

Fermat, could he but know it, might be consoled by the fact that no one has succeeded in the particular quest in which he failed. A formula that will generate prime numbers only has yet to be found.

*

The expression $2^{2^n}+1$ turned up with renewed historical importance toward the end of the eighteenth century. Anyone who has

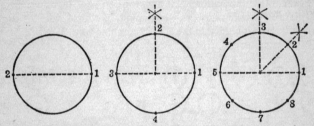

Division of the circle into 2, 4, 8, ... equal parts

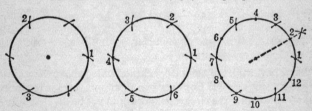

Division of the circle into 3, 6, 12, ... equal parts

FIG. 17

taken a course in plane geometry knows, as did the ancient Greeks; that the circle can be divided by means of ruler and compasses into certain numbers of equal parts. For example, to divide a circle into 2 equal parts, it is necessary only to draw a diameter. Each of the resulting semicircles can then be bisected, giving 4 equal parts; these can be bisected, giving 8 equal parts; and so on. The circle can be divided into 6 equal parts by starting at any point on the circumference and swinging successive arcs with radii equal to that of the circle. By taking every other point of division, 3 equal parts are obtained. The 6 equal parts can be doubled to 12, the 12 doubled to 24, and so on. A third method, somewhat more complicated, results in a division of the circle into 5, and, through doubling, into 10, 20, 40, . . . equal parts. The methods for 6 and 10 can be combined to give 15 equal parts. Again, the 15 can be doubled and redoubled by simply bisecting the arcs. We can state all we have been saying in the following compact way. *The circle can, by means of ruler and compasses, be divided into 3, 5, or 2^n (n any whole number) equal parts, or any combination of these such as 3×5, 3×2^n, 5×2^n, and $3 \times 5 \times 2^n$.*

For fear that this statement is a bit too compact, let us write out the first fifty whole numbers – beginning, of course, with 2 – and circle all those included in the 3, 5, 2^n combinations. We then have the following array:

②	③	④	⑤	⑥	7	⑧	9	⑩	11
⑫	13	14	⑮	⑯	17	18	19	⑳	21
22	23	㉔	25	26	27	28	29	㉚	31
㉜	33	34	35	36	37	38	39	㊵	41
42	43	44	45	46	47	㊽	49	50	51

As we have said, the general conclusion stated above was known to the ancient Greeks. For over two thousand years it remained unknown whether the circle could or could not be divided into 7, 9, 11, 13, 17, 19, 21, 23, 25, . . ., or any odd number of parts not covered in the 3, 5, 2^n combinations. Note that we say *odd* number of parts. Even numbers need not be considered, because of the factor 2^n. For example, the first uncircled even number in our array is 14. Now if the circle can be divided into 7 equal parts, then a division into 14 equal parts can be obtained by simply bisecting each of the arcs. Conversely, if a division into 14 equal parts is possible, then 7 equal parts can be obtained by simply taking every other point of division.

In 1796 a young German mathematician, Gauss, settled the question once and for all. He proved that *it is possible to divide the circle into an odd number of equal parts if, and only if, the number is a prime Fermat number* – a number of the form $2^{2^n}+1$, *that is – or any combination of such numbers.* Now the only known prime Fermat numbers are those we discussed in the last section: 3, 5, 17, 257, and 65,537. Consequently the construction is possible not only for the odd numbers 3, 5, and 3×5, but also for 17, 257, 65,537, 3×17, 3×257, $3 \times 65,537$, 5×17, 5×257, and so on. But it is *not* possible for 7, 9, 11, 13, 19, 21, 23, 25, . . . Thus, in our array of the first fifty numbers, we can now circle 17, 2×17 or 34, and 3×17 or 51, but none of the others.

The names of Fermat, Euler, and Gauss, mentioned in connection with the last two problems, are among the great names in the history of mathematics. It is interesting to note that Fermat was an amateur in the subject. He was by profession a judge – a councillor for many years in the local parliament of the city of Toulouse.

*

Powers of 2 are involved in still another matter of historical interest. The early Greeks classified numbers not only as even or odd, prime or composite, but also as perfect, excessive, or defective. Consider the number 12. Its divisors, apart from 12, are 1, 2, 3, 4, and 6. And the sum of these divisors is 16, which is greater than 12, the number itself. The number 12 is therefore said to be defective. 14, on the other hand, is excessive, for the sum of its divisors – 1, 2, and 7 – is 10, which is less than the number itself. But 6 is a perfect number, for the sum of its divisors – 1, 2, and 3 – is equal to the number itself. The next perfect number is 28. Its divisors are 1, 2, 4, 7, and 14. No odd numbers have ever been found to be perfect, but it has never been proved that there are none.

Euclid proved that any number of the form $2^{n-1}(2^n-1)$, where n is a whole number, is perfect provided 2^n-1 is prime. The only values of n for which it is *known* that 2^n-1 is prime are the following twelve: 2, 3, 5, 7, 13, 17, 19, 31, 61, 89, 107, and 127. Hence only twelve perfect numbers are known. The first six of these are 6, 28, 496, 8128, 33,550,336, and 8,589,869,056. The difficulty in working with the larger ones is easily seen if we note that the last of them is given by $2^{126}(2^{127}-1)$. The second factor, $2^{127}-1$, is the 39-digit number which appeared in connection with the two chess-boards (page 36). This, multiplied by 2^{126}, gives a number of 77 digits![3]

THE BINARY NUMBER SYSTEM

An an introduction to the present section, let us consider a case of fallacious reasoning on the part of a beginner in arithmetic. He started with the two identities

$$9+8+7+6+5+4+3+2+1 = 45,$$
$$1+2+3+4+5+6+7+8+9 = 45,$$

and subtracted the second identity from the first. The difference of the right-hand sides was 0. Beginning to the left of the equality signs, he set out to subtract 9 from 1. To do so, he had to borrow 1 from the 2 and subtract 9 from 11. This gave 2. He then subtracted 8 from 1 (which was 2 before 1 was borrowed in the last step), and to do so, he borrowed 1 from the 3 and subtracted 8 from 11, which gave 3. Then he subtracted 7 from 2, borrowing 1 from the 4, and so on, proceeding always to the next step on the left. These operations, when completed, resulted in the conclusion that

$$8+6+4+1+9+7+5+3+2 = 0,$$

or that $45 = 0$. Where did our beginner go wrong? He seems to have done nothing different from what we do in subtracting say, 189 from 321. Let us look at this illustration in more detail.

$$
\begin{array}{r}
321 \\
-189 \\
\hline
132
\end{array}
$$

Here, in subtracting the first digits at the right, we borrow 1 from the 2 and subtract 9 from 11. But do we really borrow 1? We do not. When we write the number 321, we do not mean $3+2+1$, but $3.100+2.10+1$, or $3.10^2+2.10^1+1.10^0$. (From this point on we shall use the dot, ., rather than the cross, ×, to signify multiplication.) Again, 57,289 means $5.10^4+7.10^3+2.10^2+8.10^1+9.10^0$. So when we *think* 'borrow 1 from the 2', we *actually* borrow 1.10^1 from 2.10^1. And in the next step, when we borrow 1 from 3 and subtract 8 from 11, we actually borrow 1.10^2 from 3.10^2 and subtract 80 from 110, which yields 30, or 3.10^1, which is the meaning of the '3' in the result.

This matter of 'positional notation' – the convention whereby the significance of a digit is indicated by means of its position in the written number – was developed by the Hindus about the beginning of the sixth century A.D. It was one of the greatest advances ever made in mathematics. If you do not believe this, just try multiplying together two large numbers expressed in Roman numerals! It may be worth noting in this connection that the numerals generally called 'Arabic' were actually an invention of the Hindus.[4]

Probably our number system is based on the number 10 because of the fact that man has ten fingers, which in early times he used (and still does use) as an aid in counting. There is really no reason why some number other than 10 should not be used as a base. It has been suggested, for example, that the base 12 be adopted, since 12 is divisible by 2, 3, 4, and 6, whereas 10 is divisible only by 2 and 5. Arithmetical calculation would be simpler with the base having the greater number of divisors. In the denary, or decimal, system (base 10) we use the ten digits 0, 1, 2, 3, 4, 5, 6, 7, 8, and 9. In the duodecimal system (base 12) we should have to invent symbols to designate the tenth and eleventh digits.

Would it not, someone may ask, be simpler to use a smaller base – one that would require fewer digits? Now the binary system (base 2) requires only the digits 0 and 1. Let us see what some of our denary numbers would look like in the binary scale.

$1 =$	$1.2^0 =$	1
$2 =$	$1.2^1+0.2^0 =$	10
$3 =$	$1.2^1+1.2^0 =$	11
$4 =$	$1.2^2+0.2^1+0.2^0 =$	100
$5 =$	$1.2^2+0.2^1+1.2^0 =$	101
$6 =$	$1.2^2+1.2^1+0.2^0 =$	110
$7 =$	$1.2^2+1.2^1+1.2^0 =$	111
$8 =$	$1.2^3+0.2^2+0.2^1+0.2^0 =$	1,000
$9 =$	$1.2^3+0.2^2+0.2^1+1.2^0 =$	1,001
$10 =$	$1.2^3+0.2^2+1.2^1+0.2^0 =$	1,010
$11 =$	$1.2^3+0.2^2+1.2^1+1.2^0 =$	1,011
$12 =$	$1.2^3+1.2^2+0.2^1+0.2^0 =$	1,100

$$13 = \qquad 1.2^3 + 1.2^2 + 0.2^1 + 1.2^0 = \qquad 1,101$$
$$14 = \qquad 1.2^3 + 1.2^2 + 1.2^1 + 0.2^0 = \qquad 1,110$$
$$15 = \qquad 1.2^3 + 1.2^2 + 1.2^1 + 1.2^0 = \qquad 1,111$$
$$16 = \quad 1.2^4 + 0.2^3 + 0.2^2 + 0.2^1 + 0.2^0 = \quad 10,000$$
$$\cdot\cdot \qquad\qquad \cdot\cdot\cdot \qquad\qquad \cdot\cdot\cdot$$
$$50 = \quad 1.2^5 + 1.2^4 + 0.2^3 + 0.2^2 + 1.2^1 + 0.2^0 = \quad 110,010$$
$$\cdot\cdot\cdot \qquad\qquad \cdot\cdot\cdot \qquad\qquad \cdot\cdot\cdot$$
$$100 = 1.2^6 + 1.2^5 + 0.2^4 + 0.2^3 + 1.2^2 + 0.2^1 + 0.2^0 = 1,100,100$$

It is evident at once that the disadvantage of this scale lies in the fact that it is laborious to write out a number even as small as our 100 – the first *three*-digit number in the denary scale requires the use of *seven* digits in the binary scale.

Some of us are probably wondering why we have gone into this matter anyway – the good old denary system we were brought up on seems to have advantages enough. Of what real use is a system such as the one with the base 2? We shall try to answer this question with two examples – one involving a method of calculation and the other a game.

*

A type of multiplication actually in use in the past requires no knowledge of the usual twelve multiplication tables other than that of the table of 2. Let us, for example, multiply 49 by 85 by this method. Write down 49 at the head of one column and 85 at the head of another. Divide 49 by 2 and multiply 85 by 2, writing the results below the original numbers. Continue dividing by 2 in the first column and multiplying by 2 in the second. When an odd number is divided by 2, throw away the remainder – this, strangely enough, leads to no errors. Stop when 1 is reached in the first column. The result is as follows:

(Divide by 2)	(Multiply by 2)
49	85
24	1̶7̶0̶
12	3̶4̶0̶
6	6̶8̶0̶
3	1360
1	2720
—	
	4165

42

Now in the second column cross out all those numbers which are opposite an *even* number in the first column. Add the remaining numbers in the second column, and the correct result, 4165, is obtained.

The workings of the method are easily seen if 49 is expressed in the binary scale. For then

$$49.85 = (1.2^5 + 1.2^4 + 0.2^3 + 0.2^2 + 0.2^1 + 1.2^0).85$$
$$= (32 + 16 + 0 + 0 + 0 + 1).85$$
$$= 2720 + 1360 + 0 + 0 + 0 + 85$$
$$= 4165.$$

Since 2^3, 2^2, and 2^1 do not appear in the binary representation for 49, 85 multiplied by 2^3 (680), by 2^2 (340), and by 2^1 (170) are not among the numbers to be added in the second column.

*

That the binary system can be used to definite financial advantage is shown in the following story. A poor young American student, brilliant in mathematics but inexperienced in the ways of the world, had saved up enough money to spend a year studying abroad. On the boat trip to Europe he fell in with a group of professional gamblers who in one evening of poker cleaned him of almost all his money. The next evening he again ran into the group, and another poker session was suggested. The young man admitted modestly that he guessed he didn't know the game well enough. Perhaps the gentlemen would care to play a somewhat different game? The gamblers agreed to this readily, counting on their cleverness and their ability to cheat at almost anything. The young man laid out on the table a number of heaps of matches.

'Now,' said he to one of the men, 'you may pick up as many of the matches of any one heap as you wish, from one match to all of the matches in that heap. I shall then do the same. We continue, playing alternately, until all of the matches are gone. Whoever has to pick up the last match loses the game.'

The rest of the story is easily imagined. The stakes were high, and by the end of the evening the young man had not only won back all his own money, but had made enough to spend several years abroad. As a matter of fact, he was still there when last heard from!

It takes a little time to explain how to force a win at this game, but some of us may want to see it through. Let's call the two players A and B and look at a few winning combinations towards the end of the game. If A can succeed in forcing B to draw from any one of the four situations shown in the diagram, he will win.

	Case 1	Case 2	Case 3	Case 4
1st pile	/ /	/ / /	/ / /	/ /
2nd pile	/ /	/ / /	/ /	/ /
3rd pile			/	/
4th pile				/

Case 1. (*a*) If B takes 1 match from the first pile, A takes all of the second pile, leaving B to pick up the last match. (*b*) If B takes all of the first pile, A takes one match from the second pile, and again B picks up the last.

Case 2 (*a*) If B takes 1 from the first pile, A takes 1 from the second and proceeds as in the first case. (*b*) If B takes 2 from the first pile, A takes all of the second pile, and B picks up the last. (*c*) If B takes all of the first pile, A takes all but one of the second.

Case 3. (*a*) If B takes 1 from the first pile, A takes the single match in the third pile and proceeds as in the first case. (*b*) If B takes 2 from the first pile, A takes 1 from the second pile. Then B takes 1, A takes 1, and B picks up the last. (*c*) If B takes all of the first pile, A takes all of the second. (*d*) If B takes 1 from the second pile, A takes 2 from the first pile. Then each take 1 and B takes the last. (*e*) If B takes all of the second pile, A takes all of the first pile. (*f*) If B takes the single match in the third pile, A takes 1 from the first pile and proceeds as in the first case.

Case 4. (*a*) If B takes 1 from either of the first two piles, A takes all of the other of these piles. Then each takes 1 and B takes the last. (*b*) If B takes all of either of the first two piles, A takes 1 from the other of these piles. Then each takes 1 and B takes the last. (*c*) If B takes either of the single matches, A takes the other and proceeds as in the first case.

These cases evidently do not represent all possible finishes to the game, but will do for our purposes. Suppose now that we replace each pile of matches in the diagram by the number of matches in that pile, expressing the number in the binary scale. Recall from the table on page 41 that 1 is written as $1 . 2^0 = 1$, 2 as $1 . 2^1 + 0 . 2^0 = 10$, and 3 as $1 . 2^1 + 1 . 2^0 = 11$. Then the diagram for the four cases considered becomes

	Case 1	Case 2	Case 3	Case 4
1st pile	10	11	11	10
2nd pile	10	11	10	10
3rd pile			1	1
4th pile				1
	20	22	22	22

In each of the four cases the sum of the digits in each column has been written at the bottom. We note that the digits in each sum are *even* numbers, 0 (which is called 'even' in mathematics) or 2 – never an odd number such as 1 or 3. Herein lies the secret of the game. The explanation will be clearer if we introduce the term 'coefficient'. The number 567 in the denary system means $5.10^2 + 6.10^1 + 7.10^0$. Here 7 is said to be the coefficient of 10^0, 6 the coefficient of 10^1, and 5 the coefficient of 10^2. Similarly, in the binary number 101, or $1.2^2 + 0.2^1 + 1.2^0$, the coefficient of 2^0 is 1, that of 2^1 is 0, and that of 2^2 is 1.

Now if A knows the game and B does not, A can force a win at the very outset of the game in the following manner. He expresses in the binary scale the number of matches in each pile and adds all the coefficients of 2^0, of 2^1, of 2^2, . . ., of as high a power of 2 as appears in any of the numbers. He then removes as many matches from some pile or other as is necessary to leave the sum of the coefficients of each power of 2 an *even* number. When B draws, he is bound to upset such an arrangement, and A repeats the process. The only exception to the rule is this: A must never leave an even number of piles containing only one match each.

In order to fix our ideas, let us work through one sample game completely. Suppose there are four piles, with 6 matches in the first pile, 5 in the second and third, and 3 in the fourth. A is to draw first. The set-up is shown in the diagram, with, at the right, the number in each pile expressed in the binary scale, together with the sum of the coefficients of the various powers of 2.

/ / / / / /	/ / / / /	/ / / / /	/ / /	110
				101
				101
				11
				323

Since the sums of the coefficients of 2^0 and 2^2 are odd, A must draw 3 matches from the first pile, leaving the arrangement

				11
/ / /	/ / / / /	/ / / / /	/ / /	101
				101
				11
				—
				224

in which the sums of the coefficients of all the powers of 2 are even. Suppose B draws 4 matches from the second pile. Then the arrangement is

				11
				1
/ / /	/	/ / / / /	/ / /	101
				11
				—
				124

A's move is then to draw 4 matches from the third pile, leaving

			11
			1
/ / /	/	/ / /	1
			11
			—
			24

Next suppose B takes all of the first pile, leaving the arrangement

		1
		1
/	/ / / /	11
		—
		13

Now if A played according to rule, he would take all of the matches in the last pile. But this play would leave two piles of one match each – the exceptional case to be avoided. His correct play is to remove 2 from the last pile, leaving an *odd* number of piles of one each. A then goes on to win. The exceptional case is not difficult to avoid, for it can occur only late in the game at a time when the end can easily be seen by using common sense alone.

This game pays large dividends in amusement for the small investment of time required to learn to express numbers in the binary scale and to add the coefficients rapidly.[5]

MIND-READING TRICKS

The ability of a 'mind-reader' to determine a number selected by someone in his audience is of the nature of a paradox to most people. We conclude this chapter with a few examples of tricks of this sort and shall show that they are based upon fairly simple arithmetical operations. Anyone interested in studying the subject further can find ample material elsewhere.[6]

*

The mind-reader (M) asks a man in his audience (A) to think of a number, multiply it by 5, add 6, multiply by 4, add 9, multiply by 5, and state the result.

A chooses the number 12, calculates successively 60, 66, 264, 273, 1365, and announces the last number.

M subtracts 165 from this result, gets 1200, knocks off the two zeros, and tells A that 12 was the number he thought of.

The trick is easily seen if put in algebraical symbols. If the number A chooses is a, then the successive operations yield $5a$, $5a+6$, $20a+24$, $20a+33$, and $100a+165$. When M is told this number, it is evident that he can determine a if he subtracts 165 and divides by 100 – or cancels the last two digits, which are always zero.

*

If M desires to tell A the result without asking any questions, he must so arrange the various operations that the original number thought of drops out. Here is an example in which three unknown numbers are introduced and done away with.

$M:$ Think of a number. Add 10. Multiply by 2. Add the number of pence in your pocket. Multiply by 4. Add 20. Add 4 times your age in years. Divide by 2. Subtract twice the number of pence in your pocket. Subtract 10. Divide by 2. Subtract your age in years. Divide by 2. Subtract the original number you thought of.

[A, who chooses the number 7, has 30 pence in his pocket, and is 20 years old, thinks: 7, 17, 34, 64, 256, 276, 356, 178, 118, 108, 54, 34, 17, 10.]

$M:$ Your result is 10, is it not?

A: Right!

In this case, if we denote *A*'s original number by *a*, the number of pence in his pocket by *b*, and his age in years by *c*, the successive operations give a, $a+10$, $2a+20$, $2a+20+b$, $8a+80+4b$, $8a+100+4b+4c$, $4a+50+2b+2c$, $4a+50+2c$, $4a+40+2c$, $2a+20+c$, $2a+20$, $a+10$, 10. Problems of this type can be set up in any number of ways.

*

Many tricks of the kind we are discussing are based upon the principle of positional notation. Consider the following:

M: Throw three dice and note the three numbers which appear. Operate on these numbers as follows: multiply the number on the first dice by 2, add 5, multiply by 5, add the number on the second dice, multiply by 10, add the number on the third dice, and state the result.

[*A* throws a 2, a 3, and a 4, and thinks: 4, 9, 45, 48, 480, 484.]
A: 484.

[*M* subtracts 250 and gets 234.] *M:* The numbers thrown were 2, 3, and 4, were they not?

A: Right!

More generally, suppose the numbers thrown are *a*, *b*, and *c* respectively. Then the specified operations give, successively, $2a$, $2a+5$, $10a+25$, $10a+b+25$, $100a+10b+250$, $100a+10b+c+250$. If 250 be subtracted from this number, the result is $100a+10b+c$, or $a.10^2+b.10^1+c.10^0$, so that the digits which appear in the final number are *a*, *b*, and *c*.

*

Another trick, based on positional notation, enables the mind-reader to tell a person his age and the number of pence he has in his pocket.

M: Multiply your age by 2, add 5, multiply the result by 50, add the number of pence in your pocket (less than 100), subtract the number of days in a year, and tell me the result.

[*A*, who is 35 years old and has 76 pence in his pocket, thinks: 70, 75, 3750, 3826, 3461.] *A:* 3461.

[*M* adds 115 to this number and gets 3576.] *M:* Your age is 35 and you have 76 pence.

A: Right!

Suppose that *A*'s age is *a*, and that the number of pence he has in his pocket is *b*. Then the operations specified by *M* yield, successively, $2a$, $2a+5$, $100a+250$, $100a+b+250$, and $100a+b-115$. If 115 be added to this last number, the result is $100a+b$. Now if A's age is a two-digit number, then $100a+b$ is a four-digit number. The first two of these four digits give the number a, and the last two digits the number b.

*

Here is a series of operations which always yields the same result.

M: Take any three-digit number whose first and last digits differ by more than 1, form a second number by reversing the digits, and subtract the smaller number from the larger. To the resulting number add the number formed by reversing *its* digits. Remember the result.

[*A* thinks: 853, 358, $853-358 = 495$, $495+594 = 1089$.]

M: The result is 1089, is it not?

A: Right!

That the result is always 1089 can be seen from the following general analysis:

Suppose the digits of the three-digit number are a, b, and c, where a is greater than c. Then the number itself is $a.10^2+b.10+c$, or

$$100a+10b+c.$$

The number formed by reversing the digits of this first number is

$$100c+10b+a.$$

Subtracting the second of these numbers from the first yields

$$100a-100c+0+c-a.$$

By means of subtracting 100 and adding 90 and 10, this number can be expressed as

$$100a-100c-100+90+10+c-a,$$
or $$100(a-c-1)+90+(10+c-a).$$

Reversing the digits of *this* number gives

$$100(10+c-a)+90+(a-c-1).$$

If the last two numbers are added, all a's and c's drop out, leaving

$$900+180+9,$$
or 1089.

*

This example and the next two are concerned with certain properties of the number 9.

M: Choose any three-digit number in which the first and last digits are unequal and form a second number by reversing its digits. Subtract the smaller number from the larger and tell me the first digit of the result.

[*A* thinks: 742, 247, 742−247 = 495.] *A:* The first digit is 4.

M: The other two are 9 and 5, are they not?

A: Right!

In general suppose *A*'s original number has the digits *a*, *b*, *c*, where *a* is greater than *c*. Then the number is $100a+10b+c$. Reversing the digits gives $100c+10b+a$. The difference is $99(a-c)$. It takes only a moment's reflection to verify the fact that $a-c$ must be 1, 2, 3, 4, 5, 6, 7, 8, or 9. The only possible final numbers are therefore these numbers multiplied by 99 – that is to say, 99, 198, 297, 396, 495, 594, 693, 792, 891. Now in all these numbers (save the first) the middle digit is 9 and the sum of the first and last digits is also 9. Hence if the first is known, so then are the other two.

*

Among the various important properties of the number 9 is the following, which we state without proof. *If any number is a multiple of 9, the sum of its digits is also a multiple of 9* (for example, 27, 54, 126, 234, 18,954). Let's see what use the mind-reader can make of this principle.

M: Think of a number, multiply by 10, subtract the original number, and add 54 (or any multiple of 9). In the resulting number strike out any digit except a 0, and read me the others.

[*A* thinks: 5238, 52,380, 52,380−5238 = 47,142, 47,142+54 = 47,196, 47,196.] *A:* 4, 1, 9, and 6.

[*M* adds these digits, gets 20, subtracts from the next greatest multiple of 9 – which is 27 – and gets 7.] *M:* The missing digit is 7, is it not?

A: Right!

This trick is easy to understand if symbols are used. Suppose *A* picks a three-digit number whose digits are *a*, *b*, and *c*. Then the number is $100a+10b+c$. Multiplying by 10 gives $1000a+100b+10c$. Subtracting the original number from this leaves $900a+90b+9c$. The multiple of 9 which is added can be denoted by $9k$, whereupon the final number is $900a+90b+9c+9k$. This number can be written as $9(100a+10b+c+k)$, so it is evidently a multiple of 9. It follows from the principle stated above that the sum of the digits of this number must also be a multiple of 9. Consequently the missing digit can always be determined by subtracting the sum of the others from the next greatest multiple of 9.

*

The trick we have just discussed can be made even more baffling in the following way:

M: Think of a number, subtract the sum of its digits, mix up the digits of the resulting number in any way, add 31 [*M* remembers that this number, divided by 9, leaves a remainder of 4], strike out any digit except a 0, and give me the sum of the others.

[*A* thinks: 1,234,567, 1,234,567−28 = 1,234,539, 5,923,143, 5,923,174, 5,923,174, 26.] *A:* 26.

[*M* subtracts 4 (the remainder in dividing 31 by 9), gets 22, subtracts it from 27 (the next multiple of 9) and gets 5.] *M:* The missing digit is 5, is it not?

A: Right!

M can replace the number 31 by any number he pleases, provided he remembers the remainder obtained in dividing by 9, and subtracts this from the sum of the digits which *A* gives him before he subtracts that sum from the next multiple of 9.

*

Let's look at just one more example before we go on to other matters.

M: Choose any prime number greater than 3, square it, add 17, divide by 12, and remember the remainder.

[*A* thinks: 11, 121, 138, $11\frac{6}{12}$, 6.]

M: The remainder is 6, is it not?

A: Right!

Here use is made of the fact – again stated without proof – that *any prime number greater than 3 is of the form $6n \pm 1$, where n is a whole number*. (The symbol $\pm$ means plus *or* minus.) Its square is then of the form $36n^2 \pm 12n + 1$. This number, when divided by 12, leaves a remainder of 1. Now *M* had *A* add 17, which, divided by 12, leaves a remainder of 5. The final remainder must thus be $1 + 5$, or 6.

M can vary this trick by asking *A* to add a number whose remainder, in dividing by 12, is, say, *k*. Then the final remainder will always be $1 + k$.

CHAPTER FOUR

Paradoxes in Geometry

AMONG the simplest of all geometrical paradoxes are the optical illusions, in which only the eye is fooled. Examples of this type are to be found in almost any elementary geometry book. They are used to warn the student against putting too much faith in the way a figure looks – a warning all too soon forgotten, as we shall see later in Chapter 6.

Consider the examples shown in Figure 18. Surely the line segment *BC* of diagram (a) is longer than the line segment *AB*. But no – actual measurement shows that they are equal. Similarly, in diagram (b), *AB* and *BC* are equal, as are *AC* and *BD* in (c). Again,

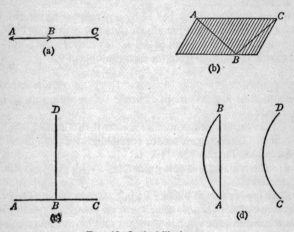

FIG. 18. Optical illusions

arcs *AB* and *CD* in (d) are equal, although the arc without the chord appears to be the longer. What of the segments *p*, *q*, and *r* in (e)? Are they segments of parallel lines? Not at all – they are parts of the same straight line. In (f), the two shaded portions have equal areas. To prove this, note that if the radius of the largest circle is taken as 5, then the inner radius of the shaded ring is 4, and the radius of the shaded circle is 3. Hence the area of the

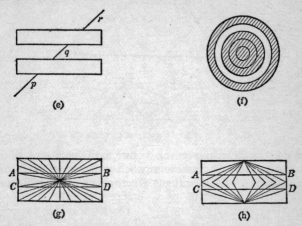

(e)

(f)

(g)

(h)

FIG. 18 (*contd.*). Optical illusions

shaded circle is $\pi r^2 = \pi . 3^2 = 9\pi$ square units, and the area of the shaded ring is $\pi . 5^2 - \pi . 4^2 = 25\pi - 16\pi = 9\pi$ square units. In (g) and (h), believe it or not, the lines AB and CD are parallel straight lines.

THE FIBONACCI SERIES

Another well-known paradox of much the same sort involves the dissection and rearrangement of a figure. It is a good example of the pitfalls of 'experimental geometry', a topic generally discussed in the early stages of any course in plane geometry. For example, the student is shown how to deduce experimentally the fact that the sum of the angles of any triangle is a straight angle, or 180°. To do so, he makes a triangle of paper or cardboard, cuts off the three angles, and rearranges them as shown in Figure 19. Let us see to what sort of contradiction this method of proof, not backed up by sound logical argument, can lead.

Suppose we take a square piece of paper and divide it into 64 small squares, as in a chessboard. We then cut it into two triangles and two trapezoids in the manner indicated in Figure 20(a) and rearrange the parts as in Figure 20(b). Now the resulting rectangle

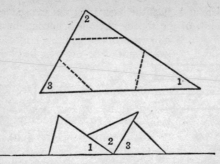

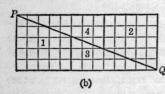

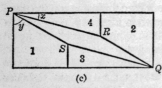

FIG. 19. The sum of the angles of a triangle is 180°

has sides which are respectively 5 units and 13 units long, so that its area is $5.13 = 65$ square units, whereas the area of the original figure was $8.8 = 64$ square units. Where *did* that additional square unit come from?

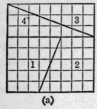

(a)

(b)

(c)

FIG. 20. The dissection and rearrangement of a square

The truth is that the edges of the parts 1, 2, 3, and 4 do not actually coincide along the diagonal PQ, but form a parallelogram $PSQR$ which is shown in exaggerated proportions in Figure 20(c). The area of this parallelogram is the elusive one square unit. The angle SPR is so small that the parallelogram is never noticed unless the cutting and rearrangement are done with great care. Indeed, it is easy for those of us who remember our trigonometry to see from the figure that $\tan x = \frac{3}{8} = \cdot 3750$ and $\tan y = \frac{5}{2} = 2\cdot 5$. Therefore $x = 20° 33'$, $y = 68° 12'$, and $\angle SPR = 90° - (20° 33' + 68° 12') = 1° 15'$.

This particular example[1] and its generalizations have engaged the attention of a number of mathematicians, Lewis Carroll among them. It is based on the relation $5.13 - 8^2 = 1$. (Recall that the dimensions of the original square were 8 by 8, and those of the resulting rectangle, 5 by 13.) The numbers 5, 8, and 13 are consecutive terms of the so-called *Fibonacci series*,

$$0, 1, 1, 2, 3, 5, 8, 13, 21, 34, 55, 89, 144, \ldots$$

Each term of this series, after the first two, is the sum of the preceding two terms. That is to say, $0 + 1 = 1$, $1 + 1 = 2$, $1 + 2 = 3$, $2 + 3 = 5$, $3 + 5 = 8$, $5 + 8 = 13$, $8 + 13 = 21$, and so on. The series is named after Fibonacci (Leonardo of Pisa), an Italian mathematician of the thirteenth century. Examples similar to ours can be constructed by using other sets of three consecutive terms, such as

$$5.2 - 3^2 = 1, \; 13.34 - 21^2 = 1, \; 34.89 - 55^2 = 1, \ldots,$$
or
$$5^2 - 3.8 = 1, \; 13^2 - 8.21 = 1, \; 34^2 - 21.55 = 1, \ldots$$

*

Although the Fibonacci series is not of any great importance in pure mathematics, the fact that it has been found to occur both in nature and in art is paradoxical enough to warrant investigation.

First let us examine the arrangement of leaves – or buds, or branches – on the stalk of a plant. Suppose we fix our attention on some leaf near the bottom of a stalk on which there is a single leaf at any one point. If we number that leaf 0 and count the leaves up the stalk until we come to one which is directly over the original one, the number we get is generally some term or other of the Fibonacci series. Again, as we work up the stalk, let us count the number of times we revolve about it. This number, too, is generally a term of the series.

If the number of revolutions is m, and if the number of leaves is n, we shall call the arrangement an 'm/n spiral'. For example, Figure 21(a) shows a $\frac{1}{2}$ spiral, as seen both from the side and from the top. The size of the stalk has been exaggerated so as to show more clearly the positions of the leaves. The arrangement in (b) can be called either a $\frac{2}{5}$ or a $\frac{3}{5}$ spiral, depending upon whether, looking down from the top, we wind about the stem in clockwise or counter-clockwise fashion. In other words, if in the first case we make 2 revolutions in counting 5 leaves, then we make $\frac{2}{5}$ revolution in passing from one leaf to the next. Consequently we must make $\frac{2}{3}$ revolution between leaves if we wind in the other direction. To fix our ideas, we shall agree to take the longer path and call this a $\frac{2}{3}$ spiral. Then the arrangement shown in (c) is a $\frac{5}{8}$ – not a $\frac{3}{8}$

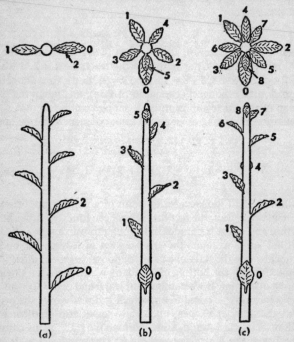

FIG. 21. The arrangement of leaves on a stalk

– spiral. Similar arrangements can be observed in a wide variety of plant growth – in pine cones, in the petals of a flower, in the leaves of a head of lettuce, and in the layers of an onion, to name but a few examples.[2]

Note that the ratios with which we have been working – $\frac{1}{2}$, $\frac{2}{3}$, $\frac{5}{8}$, and so on – are *ratios of successive terms of the Fibonacci series*. In order to study the significance of these ratios, we must turn back a couple of thousand years to the ancient Greek geometers. They were much interested in what they called the 'golden section', or the division of a line in mean and extreme ratio. The point B of

Figure 22 is said to divide the line *AC* in mean and extreme ratio if the ratio of the shorter segment to the longer is equal to the ratio of the longer segment to the whole line – that is, if AB/BC

Fig. 22, The golden section: $AB/BC = BC/AC$

$=BC/AC$. It can be shown algebraically that either of these ratios has the numerical value $(\sqrt{5}-1)/2$, which, to six decimal places, is equal to ·618034. In other words, $AB/BC=BC/AC=$·618034. Let us denote this ratio by *R*.

Now return to the Fibonacci series and consider the ratio of any term to the succeeding term. The following table gives the values of the first twelve of these ratios, calculated to six decimal places.

(1)	$1/1 =$1·000000	(2)	$1/2 =$ ·500000
(3)	$2/3 =$ ·666667	(4)	$3/5 =$ ·600000
(5)	$5/8 =$ ·625000	(6)	$8/13 =$ ·615385
(7)	$13/21 =$ ·619048	(8)	$21/34 =$ ·617647
(9)	$34/55 =$ ·618182	(10)	$55/89 =$ ·617978
(11)	$89/144 =$ ·618056	(12)	$144/233 =$ ·618026

$$\downarrow \qquad\qquad\qquad\qquad \downarrow$$
$$\text{·618034} \qquad\qquad\qquad \text{·618034}$$

The arrows indicate what is intuitively evident – that the column on the left consists of numbers which approach *R* through values greater than *R*, while the column on the right consists of numbers which approach *R* through values less than *R*. Consequently *the Fibonacci series provides a sequence of whole numbers whose successive ratios are more and more nearly equal to the ratio R of the golden section.*

Consider next the reactangle shown in Figure 23(a). The dimensions of this rectangle have been so chosen that the ratio of the width to the length is *R*. That is, $W/L=$ ·618034, or $W=$ ·618034*L*. If this rectangle is divided by a line into a square and a rectangle, as in diagram (b) of the same figure, the new rectangle is *again* one in which the ratio of the dimensions is *R*. Figure 24 shows the result of the continued division of each successive rectangle into a square and a rectangle, and shows also how a curve can be inscribed in the successive squares. This curve is known in mathematics as a 'logarithmic spiral'.[3] Remarkably enough, it is just the kind of spiral frequently found in the arrangements of seeds in flowers, in the shells of snails and other animals, and in certain

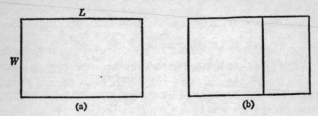

FIG. 23. The ·618034 rectangle and its division into a square and a second ·618034 rectangle

cuts of marble. (It is true that a logarithmic spiral can be inscribed in any rectangle, but the construction is not as simple as in the case discussed here. A second reason for introducing the spiral through this particular rectangle is that the rectangle itself will be mentioned shortly in another connection.)

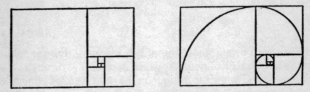

FIG. 24. Further division of the ·618034 rectangle into squares, and the inscribed logarithmic spiral

One of the best examples of the occurrence in nature of the ratio R is to be found in the head of a sunflower, shown diagrammatically in Figure 25. The seeds are distributed over the head in spirals which radiate from the centre of the head to the outside edge, unwinding in both clockwise and counter-clockwise directions. Detailed study of these spirals has resulted in the following conclusions:

(1) The spirals are logarithmic spirals.

(2) The number of clockwise spirals and the number of counter-clockwise spirals are successive terms of the Fibonacci series; and

thus the ratio of the smaller of these numbers to the larger is what appears to be nature's best possible approximation to the ratio R of the golden section.

The normal head – 5 to 6 inches in diameter – will generally have 34 spirals unwinding in one direction and 55 in the other. Smaller heads may have $\frac{21}{34}$ or $\frac{13}{21}$ combinations, and abnormally large heads have been grown with $\frac{89}{144}$ combinations. The same phenomena can be observed, although perhaps not so easily, in the heads of other flowers, such as daisies and asters.

FIG. 25. Distribution of seeds in a sunflower head

So much for the relation of the Fibonacci series to nature. What of its relation to art? It is said that psychological tests have established the fact that the rectangle most pleasing to the eye is the one shown in Figure 23 – that is, one in which the ratio of the dimensions is R. This rectangle, together with the associated logarithmic spiral, is fundamental in the technique of what has come to be called 'dynamic symmetry'. The development of the technique is chiefly the work of Jay Hambidge, who first made an intensive study of its use in the design of Greek pottery, and then extended it to sculpture, painting, architectual decoration, and even to furniture and type display.[4] Dynamic symmetry has been used extensively by a number of artists, among them George Bellows, the well-known American painter.

The apparent aesthetic appeal of dynamic symmetry is perhaps due in part to the fact that the ratio ·618034 is so universal a constant in nature. Is it because we, who are the judges of aesthetic appeal, are ourselves a part of nature? We had better leave that question to the philosopher and the psychologist, and get on with our own business.

SOME 'CIRCULAR' PARADOXES

Consider the two equal circular discs, A and B, of Figure 26. If B is kept fixed and A is rolled round B without slipping, how many revolutions will A have made about its own centre when it is back in its original position? The answer, if obtained without the aid of actual discs, is almost invariably incorrect. It is

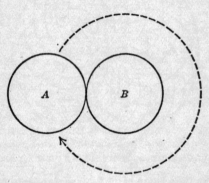

FIG. 26. Rolling a disc about an equal disc

generally argued that since the circumferences are equal, and since the circumference of A is laid out once along that of B, A must make 1 revolution about its own centre. But if the experiment is tried with, say, two coins of the same size, it will be found that A makes 2 revolutions. This fact can be shown diagrammatically as follows:

In Figure 27, let P be the extreme left-hand point of A when A is in its original position. A moment's thought will make it clear that when A has completed half its circuit about B, the arc of the shaded portion of A will have been laid out along that of the shaded portion of B, and P will again be the extreme left-hand point of A.

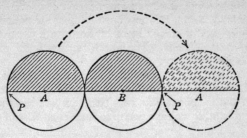

FIG. 27. The rolling disc at the half-way point

Hence *A* must have made 1 revolution about its own centre. The same argument holds for the arcs of the unshaded portions of *A* and *B* when *A* has completed the second half of its circuit about *B*.

*

Similar difficulties are encountered in the problem of a slab supported by rollers – a device frequently used in moving safes, houses, and other heavy objects.

If the circumference of each roller in Figure 28 is 1 foot, how far forward will the slab have moved when the rollers have made 1 revolution? Again the usual argument is to the effect that the distance moved must be equal to the circumference of the rollers, or 1 foot. And again the correct answer is not 1 foot, but 2 feet.

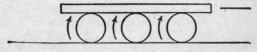

FIG. 28. The slab-and-roller problem

For suppose we resolve the motion into two parts. First think of the rollers lifted off the ground and supported at their centres. Then if the centres remain stationary, 1 revolution of the rollers will move the slab forward 1 foot. Next think of the rollers on the ground and without the slab. Then 1 revolution will carry the centres of the rollers forward 1 foot. If now we combine these two motions, it should be evident that 1 revolution of the rollers will carry the slab forward a distance of 2 feet.

In moving heavy objects by means of a slab and rollers, would it be possible to use rollers whose cross-sections are not circles, but some other sort of curve? In other words, are circles the only *curves of constant breadth*? The intuitive answer is yes; the correct answer is no.

By a curve of constant breadth we shall mean exactly what the slab-and-roller idea implies. That is to say, if such a curve is placed between and in contact with two fixed parallel lines, then it will remain in contact with the two fixed lines regardless of how it is turned.

The simplest curve of constant breadth – apart from the circle – is shown in Figure 29(a). To construct it, first draw the equilateral triangle *ABC* and denote the length of each of its sides by *r*. With *A* as centre, and with radius *r*, draw the arc *BC*. With *B* as centre, and with radius *r*, draw the arc *CA*. Finally, with *C* as centre, and with radius *r*, draw the arc *AB*. This curve can be made smooth

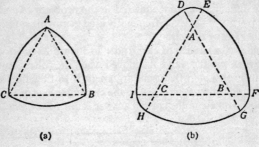

(a)   (b)

FIG. 29. Curves of constant breadth

by prolonging the sides of the triangle any distance – say *s* – as in Figure 29(b). Here the arcs *DE*, *FG*, and *HI*, with centres at *A*, *B*, and *C* respectively, are all drawn with radius *s*; and the arcs *EF*, *GH*, and *ID*, with centres at *C*, *A*, and *B* respectively, are all drawn with radius *r+s*.

In Figure 30, the second of these curves is shown placed between two fixed parallel lines. It is evident from the figure that the curve will remain in contact with the two lines regardless of how it is turned, for the distance *PQ* between the highest and lowest points of the curve is always the sum of the two constant radii, *s* and *r+s*, and so is always the same.

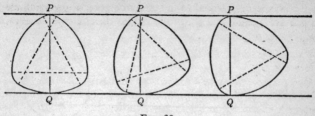

FIG. 30

It is well to note that although any roller whose cross-section is a curve of constant breadth can be used in place of a circular roller for the moving of objects on a slab, a wheel in the shape of either of the curves of Figure 29 could never be used in place of a circular cart-wheel or a circular gear. For these curves have no real centres – no point, that is, which is equidistant from all points on the curve. The circle is the only curve which has this particular property.

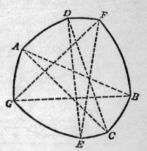

FIG. 31. An irregular curve of constant breadth

Curves of constant breadth need not be regular in shape, as were the two just examined. The irregular curve of Figure 31 is constructed as follows: with *A* as centre, and with any radius *AB*, swing arc *BC*. With *C* as centre, and with the same radius (the radius remains constant throughout), swing *AD*. With *D* as centre, swing *CE*. With *E* as centre, swing *DF*. With *F* as centre, swing *EG*. With *B* as centre, swing *AG*. (*G* is the point of intersection of the last two arcs.) Finally, with *G* as centre, swing *FB*. This curve has corner points which can be rounded off by extending the lines *AB*, *AC*, and the like, as was done in the transition from diagrams (a) to (b) in Figure 29.[5]

*

The large circle of Figure 32 has made 1 revolution in rolling, without slipping, along the straight line from *P* to *Q*. The distance *PQ* is thus equal to the circumference of the large circle. But the small circle, fixed to the large one, has also made 1 revolution, so that the distance *RS* is equal to the circumference of the small circle. Since *RS* is equal to *PQ*, it follows that *the circumferences of the two circles are equal!*

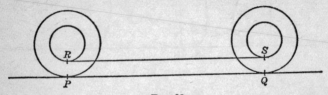

FIG. 32

This puzzling contradiction, which dates back to the seventeenth century,[6] can be explained by the fact that although the large circle rolls without slipping, the small one 'slips' in a certain sense. This behaviour can be made clear by thinking of the circles as wheels, securely fastened together, and running on tracks as shown in Figure 33. If track *b* is lowered so that it does not touch wheel *B*, then 1 revolution of the system on track *a* will carry the common

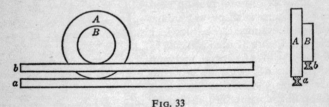

FIG. 33

centre forward a distance equal to the circumference of *A*. If, on the other hand, track *a* is lowered so that it does not touch wheel *A*, then 1 revolution of the system on track *b* will carry the common centre forward a distance equal to the circumference of *B*. Finally, suppose that each wheel rests on its corresponding track. Now the circumferences of the two wheels are certainly not equal. Consequently, if wheel *A* rolls on track *a* without slipping, there must be some slipping between wheel *B* and track *b*. And if *B* rolls on *b* without slipping, there must be some slipping between *A* and *a*. It follows that if each wheel were geared to its tracks, motion would be impossible.

A further explanation of the paradox involves the notion of a curve called the 'cycloid'. This curve, shown in Figure 34, is the path traced by a fixed point M on the circumference of a circle as the circle rolls, without slipping, along a straight line.

FIG. 34. The cycloid

A fixed point N inside the circle describes what is known as a 'curtate cycloid', or sometimes a 'trochoid'.

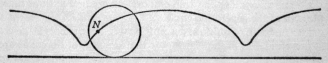

FIG. 35. The curtate cycloid

Returning to the problem of the two unequal circles, think of the motion of a fixed point M on the circumference of the large circle, and that of a corresponding point N on the circumference of the small circle. As the large circle rolls from P to Q, M describes a cycloid, and N a curtate cycloid. A glance at Figure 36 makes it

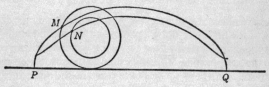

FIG. 36

evident that although each wheel makes only 1 revolution, the point M travels considerably farther than the point N. Only the common centre of the circles travels a distance equal to the straight line PQ.

*

The cycloid has a number of remarkable properties, of which we note the following three.

(1) The length of one arch of a cycloid is equal to the perimeter of a square circumscribed about the generating circle.

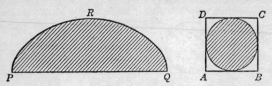

FIG. 37. The cycloidal arch *PRQ* is equal in length to the perimeter of the square *ABCD*; the area of the shaded region under *PRQ* is three times that of the shaded circle

(2) The area under one arch of a cycloid is equal to three times the area of the generating circle.

(3) The 'path of quickest descent' between two points is the arc of a cycloid. For example, suppose that *A* and *B* of Figure 38 are two points not in the same horizontal plane, and suppose that two

FIG. 38. A cycloidal arc is the path of quickest descent

spheres are released simultaneously at *A* and allowed to roll from *A* to *B*. If the first rolls along a plane, and the second along a surface in the shape of an inverted cycloid, the second will arrive at *B* before the first, in spite of the fact that its path is longer and that it has to roll *uphill* before it gets to *B*. It can be shown further that if the plane from *A* to *B* is replaced by a curve of any other shape, the sphere which rolls along this surface will always arrive at *B* later than the one which rolls along the cycloid.

This problem of the path of quickest descent, traditionally known as the 'brachistochrone problem', was proposed to Jacob Bernoulli by his brother Johannes (see p. 178) in 1696. It was not long before the methods devised for the solution of the problem developed into what is now called the 'calculus of variations' – an important branch of mathematics dealing with all sorts of extremal problems.

A third member of the cycloid family is the 'prolate cycloid'. This curve is the path traced by a fixed point O outside the rolling circle, but attached to it. Figure 39 shows that as the circle rolls to

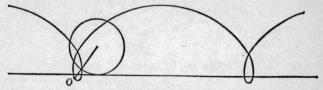

FIG. 39. The prolate cycloid

the right, the point O moves to the left during a small part of its journey. It can thus be said that no matter how fast a train is moving *forwards*, certain parts of the train – points on the flanges of the wheels – are moving *backwards*!

FIG. 40. The hypocycloid

Incidentally, a nice paradox exists in connection with the naming of the curtate and prolate cycloids. We have labelled Figures 35 and 39 in accordance with the definitions given in the *Encyclopaedia Britannica* (14th edition, 1939). But according to Webster's *New International Dictionary* (2nd edition, 1934), what we have called a curtate cycloid should be called a prolate cycloid and vice versa. In view of the fact that 'curtate' is derived from 'curtus', meaning 'short', and 'prolate' from 'prolatus', meaning 'prolonged', the terminology we have adopted would seem to be the obvious one, in spite of Webster.

One other type of cycloid also deserves mention. The 'hypocycloid' is the path of a fixed point P on the circumference of a circle which rolls around the interior of a larger, fixed circle.

If, as in Figure 41(a), the radius of the rolling circle is half that of the fixed circle, the point P simply moves back and forth along the diameter AB. Here, then, is a device by which circular motion can be transformed into straight-line motion. Figure 41(b) shows the centre C of the rolling circle attached to a revolving disc D, and a rod attached to the rolling circle at P. The rotation of the

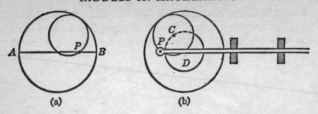

FIG. 41. Transforming circular motion into straight-line motion

disc *D* causes the small circle to roll about inside the large circle (which remains fixed), and this rotation in turn causes the rod to move back and forth in a straight line.

TOPOLOGICAL CURIOSITIES

Let us think for a moment of the shape of a quoit, or anchor-ring. Is that part which constitutes the hole inside or outside the ring? We generally avoid the wordy phrase used here and speak of 'the hole in a ring', implying, however unconsciously, that it is inside. But is the inside really inside or outside? If we go on to debate the question without first settling upon some sort of definition of 'inside' and 'outside', our argument is likely to be quite fruitless.

The problem of what constitutes the inside and the outside of a ring is the concern of the student of 'topology' or 'analysis situs' (literally, the analysis of situation, or position). Ordinary plane and solid geometry are essentially quantitative, dealing as they do with the size of things – the *lengths* of lines, the *areas* of surfaces, and the *volumes* of solids. Topology, on the other hand, is a kind of geometry which ignores sizes and concentrates on such qualitative questions as whether a certain point is *inside*, *on*, or *outside* a certain closed curve or surface.

To be more specific, consider the circle shown in Figure 42. The student of plane geometry is interested in such things as the number of inches in the circumference of the circle, or in the number of inches in the distance from the centre *O* to the point *P*, or in the number of square inches in the area of the circle. The topologist, on the other hand, is interested in this sort of question: The point *P* is inside the circle, the point *Q* on the circle, and the point *R* outside the circle. Now suppose the circle is drawn on a sheet of rubber and then stretched and distorted in any way, shape, or manner – provided it is not torn. Does P still lie inside the curve? Does Q still lie on the curve? Does R still lie outside the curve? The answer to all three of these questions is obviously yes, but this problem is an elementary one.

The science of topology is relatively young. The first systematic work in the subject appeared about the middle of the nineteenth

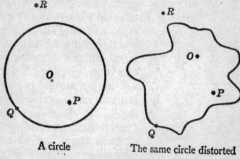

| A circle | The same circle distorted |

FIG. 42

century. But in 1736, over a hundred years earlier, Euler published the first single result of any topological consequence. Let us look at his problem.[7]

In the German town of Königsberg ran the river Pregel. In the river were two islands, connected with the mainland and with each other by seven bridges, as shown in Figure 43. A frequent topic of

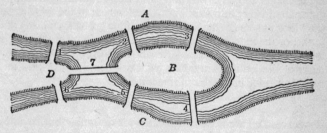

FIG. 43. The bridges of Königsberg

conversation in the town was whether or not it was possible for a person to set out for a walk from any point in the town, cross each bridge once and only once, and return to his starting-point. No one had ever found a way to do this, but on the other hand, no one had ever been able to prove that a way did not exist. Euler heard of the problem and went about its solution in a systematic manner. He noted – and here the topological method creeps in – that the problem is unchanged if the somewhat complicated figure above is replaced by the simple diagram in Figure 44. Then the original

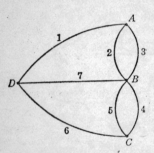

FIG. 44. The Königsberg problem simplified

problem is equivalent to this: is it possible to start at any point and trace this diagram with a pencil without lifting the pencil from the paper without retracing any portion of the diagram? Euler proved not only that this is impossible, but went on to establish additional results for diagrams of a more general nature. Incidentally, the diagram above can be traced in the manner indicated if the bridge *BD* is replaced by one from *A* to *C*.

Euler's problem can hardly be called paradoxical, but there are two reasons why it was worth discussing. First, it gives us some idea of the topological method, whereby a complicated diagram is replaced by a simple one. In the second place, it indicates the general nature of a topological problem – one in which the essentials are unchanged by any distortion of the figure. But now, instead of going any further into a technical development of the subject, we shall look at some of the weird and startling problems which arise in it. For the most part we shall be dealing with what we always thought were simple ideas – ideas about which our intuition has never before led us astray. Perhaps it will become clear how unreliable a guide intuition can sometimes be.

*

The curve in Figure 45 is a complicated-looking affair, but a mathematician would call it a 'simple closed curve', for it never crosses itself, and it divides the plane in which it lies into two parts, one inside the curve and the other outside. Topologically speaking, it is equivalent to a circle, for it can be transformed into one by proper stretching. Figure 46, on the other hand, shows a closed curve which is not simple. At first thought we may be tempted to say that this curve divides the plane into an inside, consisting of the regions I and X, and an outside, O. But not so fast! Let us return for a moment to Figure 45. If we start at any point or the

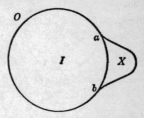

FIG. 45. A simple closed curve

FIG. 46. A closed curve that is not simple

inside, I, and follow a path which cuts the curve at *any one point*, we shall find ourselves at some point of the outside, O. This corresponds roughly to our intuitive idea of what constitutes the inside and the outside of a curve. Now, in Figure 46, if we start anywhere in I and follow a path which cuts the curve anywhere *except* between the points a and b, it is true that we find ourselves at some point of O. But what if our path cuts the curve *between a and b*, so that we arrive in X? If X is outside the curve we ought to be able to get from X to O without again crossing the curve. And if X is inside the curve, we should have been able to get from I to X without crossing the curve the first time. Hence, relative to I, the region X is neither inside nor outside the curve.

*

So far we have been working with one-dimensional curves on a two-dimensional surface. We shall now step everything up one dimension and consider a similar problem involving two-dimensional surfaces in three-dimensional space.

A sphere is a good example of a 'simple closed surface' – a surface which divides all of space into two regions, one inside the sphere and the other outside. If we start at any point of the inside, I, and follow a path which cuts the surface at any one point, we arrive at the outside. O, as in Figure 47(a).

The surface whose cross-section is shown in Figure 47(b) is

constructed by taking a hollow sphere, soldering a hollow pipe to the outside of the sphere at *A*, cutting a hole in the sphere at *B*, and here soldering the other end of the pipe to the sphere. Thus the

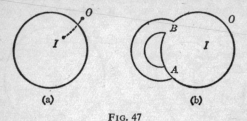

(a)   (b)

FIG. 47

pipe is closed at *A*, and opens into the sphere at *B*. The surface formed by the sphere and the pipe is certainly a closed surface, but it is no longer a simple one. For what constitutes the inside, and what the outside? If we start from any point of *I* and follow a path which cuts the sphere anywhere *except* in the circle at *A*, we arrive at *O*. But if our path cuts the sphere *in the circle* at *A*, we follow the pipe around and again find ourselves inside the sphere. And if we follow the same path in reverse order, we are still inside the sphere.

This problem is considerably more baffling than that of the curve in Figure 46. There we might at least have said that the curve has two separate insides, *I* and *X*, and one outside, *O*. But here the sphere and the pipe cannot be thought of as separate insides, for although they are separated at *A*, they run into one another at *B*. The best we can say is that the surface has an inside and an outside except for the small portion of the sphere at *A*.

Now look at the surface – called 'Klein's bottle' – shown in Figure 48. We can construct this surface by taking one end of a hollow glass tube, bending it round, inserting it through a hole in

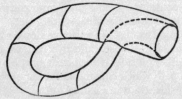

FIG. 48. 'Klein's bottle' – closed surface with no inside

its side, and welding the two open ends together. The resulting surface is a closed surface, being unbroken in the usual sense at any point. For example, Figure 49(a) shows the cross-section of an ordinary bottle. This surface is an open one, being broken at the neck. Figure 49(b), on the other hand, is a cross-section of the surface of Figure 48. This surface has no break like that at the neck of the bottle. To repeat, it is a *closed* surface. (In all cases, of course, the glass must be thought of as a true surface – one with no thickness.)

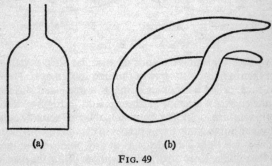

(a)   (b)

FIG. 49

Suppose we start anywhere and follow a path which cuts the surface at any one point. We can, without again cutting the surface, return to the place from which we started. In other words, no matter where we penetrate the surface, we are still outside of it. *This closed surface therefore has no inside whatever!*[8]

*

Most of the surfaces met with in everyday life are 'bilateral', or two-sided. A sheet of paper, for example, has two sides. If a fly were placed on one side, he could get to a point on the other side only by cutting through the paper – how a fly could do this is beside the point – or by going over the edge. A sphere is a closed bilateral surface. The fly could crawl all over the outside, and could get to the inside only by going through the surface. But the closed surface of Figure 48 is 'unilateral', or one-sided. A moment's thought will make it clear that the fly could crawl from any one point to any other point without the inconvenience of cutting through.

Let us consider a simpler example of a unilateral surface – one which is somewhat easier to construct. First take a long, narrow, rectangular strip of paper, and paste the ends together as shown in Figure 50. The result is a cylindrical surface which has two sides and two edges. We shall refer to this strip as S_1.

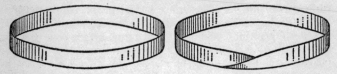

FIG. 50. An ordinary cylindrical strip (S_1)   FIG. 51. The Möbius strip (S_2)

Now, before pasting the ends of a similar strip together, give one of them a half-twist – a twist through 180°, that is. The resulting surface, called a 'Möbius strip', is a one-sided surface with only one edge. An attempt has been made, in Figure 51, to show what this surface looks like, but you had better actually construct one if you want to study its properties in detail. To convince yourself that it has only one side, start at any point and draw a line down the middle. Keep on drawing, without lifting the pencil from the paper, until you return to the point from which you started. You will find that the single line has completely traversed what constituted, *before* the ends were pasted together, the *two* sides of the original rectangular piece of paper. And to convince yourself that the Möbius strip has only one edge, start at any point of the edge and follow it round, without crossing the paper until you

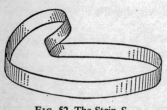

FIG. 52. The Strip S_3

are back where you started. Again, you will find that you have completely traversed what constituted, *before* the ends were pasted together, the *two* long edges of the original rectangular piece of paper. We shall, for convenience, call the Möbius strip S_2.

Finally, if one of the ends is turned through a full twist (through 360°) before pasting, the resulting surface, like S_1, has two sides and two edges. We shall refer to this strip, illustrated in Figure 52, as S_3.

And now get out your scissors, for we have more to do. Suppose we cut the bilateral strip S_1 along a line midway between the edges. It is not difficult to see that we obtain two separate strips, identical with the original one except that they are only half as wide. But what if we cut the Möbius strip S_2 in the same manner? Anyone who can predict the result before actually carrying out the experiment must have better than average intuitive powers. For the result is a *single* strip – not two – twice as long and half as wide as the original one. Furthermore, it is no longer unilateral, but is a bilateral strip of the type S_3. And what if this strip S_3 is cut down the middle? Here the result consists of two interlocked surfaces of the type S_3, each of them equal in length to the strip from which they were cut, and half as wide.[9]

For a final surprise, take a new strip S_2 – the unilateral Möbius strip – and cut it along a line which runs parallel to the edge and a third of the width of the strip from the edge. Keep cutting until you are back at the starting-point. The result? Two interlocked surfaces, one equal in length to the original strip, the other twice as long. The shorter one is again of type S_2, The longer one of type S_3.

Try the experiments described and invent variations of your own. Predicting the results will give your intuition good practice.

*

Much of the early work in topology was concerned with the study of knots. As a matter of fact, anyone who has ever played with those puzzles consisting of interlocked wires, nails, rings, or strings can claim to have been, in a small way, a topologist.

When is a closed curve knotted and when is it *not* knotted? A piece of string with the ends tied together will do nicely for a closed

(a)   (b)

FIG. 53. Knotted or not knotted?

curve. Then the curve, or string, is not knotted if it can be transformed into a single simple closed curve (see pp. 70–1) without cutting and retying. Otherwise it is knotted. For example, Figure 53(a) shows a string which is actually unknotted, while diagram (b) of the same figure shows one which is knotted in the simplest possible way.

Our discussion of the Möbius strip and its various relatives can be put in terms of knots as follows.

The two edges of a strip of type S_1 are neither knotted nor interlocked. This strip, cut down the middle, falls into two separate parts.

The single edge of a strip of type S_2 is not knotted. This strip, cut down the middle, becomes a single unknotted strip of type S_3.

The two edges of a strip of type S_3 are interlocked, but not knotted. When this strip is cut down the middle, it falls into two interlocked strips.

If one end of the strip is turned through three half-twists (or 540°) before pasting, the resulting surface – call it S_4 – has one side

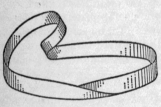

FIG. 54. The strip S_4

and one knotted edge. For if this strip is cut down the middle, a single strip is obtained, as in a strip of type S_2, but in this case the strip itself is knotted. As a matter of fact, the knot is of the type shown in Figure 53(b).

There are many examples of paradoxes which arise in the study of knots. The reader who is interested further in the subject is referred to other works.[10] We shall content ourselves with a description of only two more examples – both of them rather popular tricks.

The first concerns our friends A and B, who appear in Figure 55. They are tied together in the following way. One end of a piece of rope is tied about A's right wrist, the other about his left wrist. A second rope is passed around the first, and its ends are tied to B's wrists. How are A and B to free themselves without cutting one of the ropes? No amount of climbing in and round each other's rope will do it, but the solution is simple. B takes up a small loop near the middle of his own rope, passes it under the loop round A's right wrist – on the inside of the wrist and in the direction from elbow to hand – and slips it over A's hand. He then passes it again under the loop round A's wrist – this time on the outside of the wrist and in the direction from hand to elbow – and behold, his own rope can now be pulled free.

A   B

FIG. 55. The escape-artist's trick

The second of our two tricks has to do with the system of apparently interlocked loops and surfaces consisting of a man, his waistcoat, and his coat. We should probably say offhand that it is impossible for a man to take off his waistcoat without first removing his coat – without, that is, slipping his arms out of his coat-sleeves. But you, or anyone else, can do it by following these directions. Unbutton the waistcoat and coat. Grasp the end of the left-hand sleeve and the lower left-hand corner of the coat firmly in the left hand and put that hand and arm through the left arm-hole of the waistcoat, from outside to inside. This operation leaves the left armhole free, and over the left shoulder. Pull the waistcoat round behind the neck. Now grasp the end of the right-hand sleeve and the lower right-hand corner of the coat firmly in the right hand and put that hand and arm through the left armhole, again from outside to inside. This operation leaves the waistcoat attached to the body only by the right arm, which is now through both armholes of the waistcoat. Finally, pass the waistcoat down the inside of the right coat-sleeve and out at the end.

Use an old suit!

*

A number of topologists have spent a great deal of time and energy on the 'four-colour problem'. Experience has taught mapmakers, both amateur and professional, that only four colours are needed to distinguish between the different countries of a plane or spherical map. Let us turn cartographers for a moment and look at a few examples of plane maps.

A piece of land occupied by one, two, three, or four countries can obviously be taken care of with four colours or fewer. Perhaps we had better put one possible misunderstanding out of the way before it arises. It may be argued that we need seven colours for the map in Figure 56. But one of the conditions of the problem is that two countries may be coloured the same if they touch only at one point – not, however, if they touch along a line. Thus in the map under consideration we can do with three colours, indicated in the figure by three distinctive shadings.

Now consider Figure 57, which shows four countries each of which touches the other three. It is evident at once that four colours are necessary for the colouring of this map. No one, however, has yet been able to draw a map of *five* countries, each of which touches the other *four*.

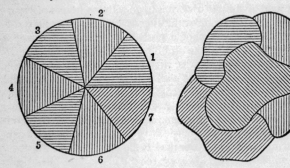

FIG. 56. Three colours suffice for this map of seven countries

FIG. 57. Four colours are necessary for this map of four countries

Here is an excellent example of what are known as *necessary and sufficient* conditions in a mathematical problem. The map of Figure 57 furnishes proof of the fact that *four colours are necessary*. Yet the mere fact that no one has ever found a map for which four colours are not sufficient does *not* prove that *four colours are sufficient*.

Although it is *suspected* that four colours are sufficient, the best result which has been *proved* to date is that five are sufficient. It is remarkable that although the problem has been solved only in part for such simple surfaces as the plane, and the sphere, it has been solved completely for much more complicated surfaces. For example, it has been proved that seven colours are both necessary *and* sufficient for the colouring of a map on a 'torus' – a quoit, or anchor-ring. An attempt is made in Figure 58 to show a torus upon which seven regions, each of which touches the other six, have been laid out.[11]

The four-colour problem has fired the imagination and enthusiasm of many a mathematician. Scarcely a month goes by but that some mathematical journal or other carries an article either on the

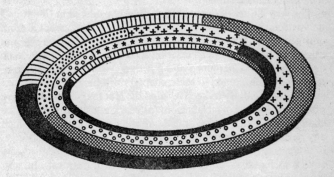

FIG. 58. The 'seven-colour problem' of the torus

problem itself, or on problems which have arisen out of it. The real problem – that of the sufficiency of four colours for a plane or spherical map is still an open question.

*

We must not conclude, from what we have seen in our brief excursion into topology, that it is a subject made up entirely of useless, though interesting, games and puzzles. The introduction of topological methods has brought about startling advances not only in other branches of mathematics but in physics and chemistry as well. And to mention only one of the applications of topology to industry, the Bell Telephone Laboratories have found a use for it in the classification of electrical networks. No one dares prophesy the ultimate usefulness of topology – it is as yet too young a science.

CHAPTER FIVE

Algebraical Fallacies

MOST of the paradoxes of the previous three chapters were paradoxical in that they appeared to be false – or at least highly improbable – yet were actually true. In this chapter and the next we shall consider some results in algebra and geometry which appear to be true, yet which are actually false. Paradoxes of this type might better be called 'fallacies', since they are the result of fallacious logical reasoning.

The very nature of the problems to be considered in Chapters 5 and 6 necessitates the use of the more formal techniques of algebra and geometry. It may well be that a number of readers will not find these two chapters as exciting as Chapters 7, 8, and 9. Such people can skip Chapters 5 and 6 if they wish, although some of the material developed in these chapters will be used later on. At any rate, we shall try to keep in mind the fact that many of us have not seen the inside of a classroom for years, and for this reason we shall generally discuss in some detail the finer points upon which any particular argument may be based.

Whenever there are groups of fallacies which involve the same error – such as division by zero, to name one of the most common – only one example of each group will be explained in full. The remaining examples of the group will then be enumerated either without explanation, or with at most a hint or two. This leaves the satisfaction of exposing the difficulty to the reader, though the complete solution is always available in the Appendix as a last resort!

*

No doubt everyone will recall the axioms, or assumptions, which are at the foundation of the study of arithmetic and which are consequently essential to any mathematical argument having to do with numbers. We probably remember them in some such singsong fashion as this: 'Equals plus or minus equals are equal; equals multiplied or divided by equals are equal; like powers or like roots of equals are equal; things equal to the same thing are equal to each other'; and so on. Let us have a look at some applications – or rather misapplications – of these axioms.

Paradox 1.

$$1 \text{ cat has 4 legs;} \tag{1}$$
$$\text{no (i.e., 0) cat has 3 legs.} \tag{2}$$

Adding the 'equals' (1) and the 'equals' (2), we conclude that 1 cat has 7 legs.

Paradox 2.

$$2 \text{ pounds} = 32 \text{ ounces;} \tag{1}$$
$$\tfrac{1}{2} \text{ pound} = 8 \text{ ounces.} \tag{2}$$

Multiplying the equals (1) by the equals (2), we obtain 1 pound $= 256$ ounces.

Paradox 3.

$$1.0 = 2.0; \tag{1}$$
$$0 = 0. \tag{2}$$

Dividing the equals (1) by the equals (2) gives $1 = 2$.

Paradox 4.

$$(-a)^2 = (+a)^2,$$

since the square of a negative quantity is positive. Extracting the square root of both sides, we have $-a = +a$.

Paradox 5.

$$\tfrac{1}{4} \text{ dollar} = 25 \text{ cents.}$$

Extracting the square root of both sides of this expression gives

$$\sqrt{\tfrac{1}{4}} \text{ dollar} = \sqrt{25} \text{ cents,}$$

or

$$\tfrac{1}{2} \text{ dollar} = 5 \text{ cents.}$$

Paradox 6. In attempting to solve the system of two equations in two unknowns:

$$\begin{cases} x + y = 1, \\ x + y = 2, \end{cases}$$

we are forced to the conclusion that, since 1 and 2 are equal to the same thing, they must be equal to each other – that is, $1 = 2$.

Where are the errors? Paradox 1 is too obvious to spend any time on. As a matter of fact, we had to stretch our imagination a little in order to get 'equals' out of the statements (1) and (2).

In Paradoxes 2 and 5 we performed the operations of multiplication and root extraction only on the *numbers*, and not on the *units* involved. Our conclusion in Paradox 2, for example, should have been

$$1 \text{ (pound)}^2 = 256 \text{ (ounces)}^2.$$

Now a 'square pound' is rather a difficult thing to visualize. It would be clearer if we used feet and inches. Our argument would then run as follows:

$$2 \text{ feet} = 24 \text{ inches};\tag{1}$$
$$\tfrac{1}{2} \text{ foot} = 6 \text{ inches}.\tag{2}$$

Therefore 1 square foot = 144 square inches – a result which is evidently correct.

Paradox 3 reminds us rather forcibly of the fact that the axiom concerning 'equals divided by equals' carries with it a rider to the effect that the divisors shall not be zero. We shall have more to say about this point before very long.

Paradox 4 recalls another item which may well have been forgotten. In extracting a square root, both the positive and negative signs must be taken into consideration. That is to say, the expression in question yields the two correct identities $+a = +a$ and $-a = -a$. Here is another matter which will receive more attention later on.

Paradox 6 shows us that the axioms cannot be applied blindly to equations which are true only for certain values of the variables, or unknowns. The values of x and y for which both the equations (1) and (2) are true must be taken into account, and there are no values of x and y for which $x + y = 1$, and, at the same time, $x + y = 2$.

Another misuse of the axioms might be of some value in the following instance. Suppose a man is accused by his indignant wife of having had too much to drink. He can maintain that it is undeniably true that a glass which is half full is equivalent to a glass which is half empty. That is, $\tfrac{1}{2}$ full $= \tfrac{1}{2}$ empty. But from this it follows – by multiplying both sides by 2 – that full = empty, so that every time our friend had drunk a full glass, he had had nothing at all to drink! The chances for the success of this scheme are inversely proportional to the wife's intelligence.

*

Not only is it possible to observe the rules (after a fashion) and come out with incorrect results, but, as any teacher of mathematics will testify, it is often possible practically to annihilate the rules and still arrive at the correct conclusion. For example, it is quite true that if in the fractions $\frac{16}{64}$ and $\frac{26}{65}$ the sixes are cancelled from the numerators and denominators, the resulting fractions, $\frac{1}{4}$ and $\frac{2}{5}$, are correct. Again, the exponents in the numerator and denominator of $(1+x)^2/(1-x^2)$ can be cancelled, and the correct result, $(1+x)/(1-x)$, obtained.

The same sort of illegal cancellation was once used in the following proof in plane geometry. As in Figure 59, the points P, O, R, Q and S are so marked off on a straight line that $PO = OQ$ and

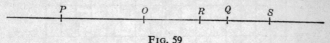

FIG. 59

$OR/OQ = OQ/OS$. It is then desired to prove that $PR/RQ = PS/QS$. This result, it was argued, must be true; for if the R's are cancelled on the left of this last equation, and the S's on the right, the identity $P/Q = P/Q$ results! But $P/Q = P/Q$ is of course meaningless, since P and Q represent points, whereas PR and RQ are magnitudes.

*

A type of examination question which has recently become popular – at least with the examiners – is this: 'If x be diminished, in what way does the fraction $1/x$ change?'[1] The student is expected to answer that the value of the fraction increases, and the value of the problem is supposed to lie in the fact that it tests the student's feeling for what is known as 'functional relationship'. The expected reply appears reasonable enough, yet it leads to a contradiction if no further restrictions are put on the values assumed by x. For suppose x runs over the decreasing sequence

$$\ldots, 5, 3, 1, -1, -3, -5, \ldots$$

Then the corresponding values of $1/x$ are

$$\ldots, \tfrac{1}{5}, \tfrac{1}{3}, 1, -1, -\tfrac{1}{3}, -\tfrac{1}{5}, \ldots$$

Now it is true enough that $\frac{1}{3}$ is greater than $\frac{1}{5}$, that 1 is greater than $\frac{1}{3}$, that $-\frac{1}{3}$ is greater than -1, and that $-\frac{1}{5}$ is greater than $-\frac{1}{3}$. But must we conclude that -1 is greater than 1?

The difficulty arises from our failure to examine carefully all possibilities before concluding that $1/x$ increases as x decreases. We probably thought of letting x run over a decreasing sequence of *positive* numbers – such as ..., 5, 4, 3, 2, 1 – in which case the value of $1/x$ does increase. It also increases if x runs over a decreasing sequence of *negative* numbers, such as $-1, -2, -3, -4, -5, \ldots$ These facts can be verified by a glance at the graph of $1/x$, shown in Figure 60. But the figure also shows that we cannot

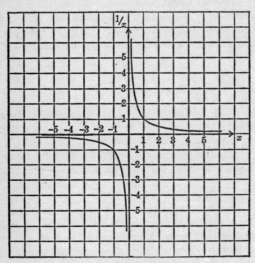

FIG. 60. The graph of $1/x$

conclude that $1/x$ increases as x runs over a decreasing sequence of *both* positive and negative numbers – there is a gap in the curve as x, in decreasing, passes through the value 0.

The same sort of negligence in defining carefully the range of permissible values which the variable or variables can assume – the 'domain of definition', as it is frequently called – leads again to the paradoxical result that -1 is greater than 1 in the following instance.[2]

Consider the proportion $a/b = c/d$. It seems reasonable to assert that if the numerator of the first fraction is greater than that of the second, then the denominator of the first fraction must be greater than that of the second. That is, if a is greater than c, then b is greater than d (as, for example, in the proportion $6/3 = 4/2$). But now suppose that $a = d = 1$, and $b = c = -1$. Then the proportion becomes $1/-1 = -1/1$, a proportion which is unquestionably valid. But since the numerator 1 is greater than the numerator -1, we must conclude that the denominator -1 is greater than the denominator 1. In other words, 1 is both greater than -1 and less than -1! Here again we should have restricted a, b, c and d to either positive numbers or negative numbers, not a mixture of both.

*

Almost everyone who has been exposed to elementary algebra has at one time or other been exposed to a proof that $2 = 1$. Such a proof is generally something of this sort:

Assume that

$$a = b \tag{1}$$

Mutliplying both sides by a,

$$a^2 = ab \tag{2}$$

Subtracting b^2 from both sides,

$$a^2 - b^2 = ab - b^2. \tag{3}$$

Factorizing both sides,

$$(a+b)(a-b) = b(a-b). \tag{4}$$

Dividing both sides by $a - b$,

$$a + b = b. \tag{5}$$

If now we take $a = b = 1$, we conclude that $2 = 1$. Or we can subtract b from both sides and conclude that a, which can be taken as *any* number, must be equal to zero. Or we can substitute b for a and conclude that any number is double itself. Our result can thus be interpreted in a number of ways, all equally ridiculous.

It may well be that we remember not only this sort of proof, but also the point at which the error occurs. It is in step (5), where we divided both sides by $a - b$. Since a and b were originally assumed to be equal, we divided both sides by zero. Now why, you may well ask, *can't* we divide by zero? The answer involves the notion of *consistency*, which we discussed briefly towards the end of Chapter 1. There it was pointed out that the mathematician asks only that his axioms lead to no such contradictions as that $2 = 1$. Let us look at this question in a little more detail.

Division in mathematics is defined by means of multiplication. To divide a by b means to find a number c such that $bx = a$, whence $x = a/b$. If $b = 0$, there are two different cases to discuss: that in which a is *not* zero, and that in which a *is* zero. Suppose we try to determine x in each of these cases. In the first we have $x = a/0$, or $0.x = a$. Now what number x, multiplied by 0, will give a, where a is any fixed number (not equal to zero), suc has 3, or -5, or $\frac{7}{8}$? Since any number multiplied by 0 is 0, there is *no* such number x. In the second case, where a is zero, we have $x = 0/0$, or $0.x = 0$. Here *any* number x will do, since, as we have said, any number multiplied by zero is zero. Now the mathematician requires that the division of a by b yield a definite, unique ('single' is meant here – not 'unusual'!) number as a result. And we have just seen that division by zero leads either to *no* number or to *any* number. Is it any wonder, then, that the mathematician has adopted the rule which some teachers refer to as the Eleventh Commandment, 'Thou shalt not divide by zero'?

Here are some other fallacies, all based on the illegal operation we have been discussing. Can you find the trouble yourself? (You will find the solutions of all numbered paradoxes – *Paradox 1, Paradox 2*, and so on – in the Appendix. This remark applies both to this chapter and to the next.)

Paradox 1. To prove that any two unequal numbers are equal.[3]
Suppose that

$$a = b + c, \tag{1}$$

where a, b, and c are positive numbers. Then inasmuch as a is equal to b plus some number, a is greater than b. Multiply both sides by $a - b$. Then

$$a^2 - ab = ab + ac - b^2 - bc. \tag{2}$$

Subtract ac from both sides:

$$a^2 - ab - ac = ab - b^2 - bc. \tag{3}$$

Factorize:

$$a(a - b - c) = b(a - b - c). \tag{4}$$

Divide both sides by $a - b - c$. Then

$$a = b. \tag{5}$$

Thus a, which was originally assumed to be greater than b, has been shown to be equal to b.

Paradox 2. To prove that all positive whole numbers are equal.[4]
By ordinary long division we have, for any value of x,

$$\frac{x-1}{x-1} = 1,$$

$$\frac{x^2-1}{x-1} = x+1,$$

$$\frac{x^3-1}{x-1} = x^2+x+1,$$

$$\frac{x^4-1}{x-1} = x^3+x^2+x+1,$$

$$\ldots = \ldots$$

$$\frac{x^n-1}{x-1} = x^{n-1}+\ldots+x^2+x+1.$$

Now in all of these identities let x have the value 1. The right-hand sides then assume the values 1, 2, 3, 4, ..., n. The left-hand sides are all the same. Consequently $1 = 2 = 3 = 4 = \ldots = n$.

Paradox 3. The following argument shows how the axioms can be violated and the corrected results still be obtained.[5]
Let x have the value 3, so that

$$x-1 = 2. \tag{1}$$

Adding 10 to the *left-hand side only*,

$$x+9 = 2. \tag{2}$$

Multiplying both sides by $x-3$,

$$x^2+6x-27 = 2x-6. \tag{3}$$

Subtracting $2x-6$ from both sides,

$$x^2+4x-21 = 0. \tag{4}$$

Factorizing,

$$(x+7)(x-3) = 0. \tag{5}$$

Dividing both sides by $x+7$,

$$x-3=0, \text{ or } x=3, \tag{6}$$

which is the value originally assigned to x.

*

Division by zero is sometimes fairly well disguised. For example, in the theory of proportions it is easy to establish the fact that if two fractions are equal, and if their numerators are equal, then their denominators are equal. That is, from $a/b=a/c$ it can be inferred that $b=c$. That this inference is not valid if $a=0$ can be seen by running through the argument in the general case. Given

$$\frac{a}{b}=\frac{a}{c}.$$

Multiply both side by bc. Then

$$ac=ab.$$

Divide both sides by a:

$$c=b.$$

But if $a=0$, then this last step involves division by zero.

Consider the following problem in this light.[6] It is desired to solve the equation

$$\frac{x+5}{x-7}-5=\frac{4x-40}{13-x}. \tag{1}$$

Combining the terms on the left-hand side,

$$\frac{x+5-5(x-7)}{x-7}=\frac{4x-40}{13-x}. \tag{2}$$

Simplifying,

$$\frac{4x-40}{7-x}=\frac{4x-40}{13-x}. \tag{3}$$

Now since the numerators in (3) are equal, so also are the denominators. That is, $7-x=13-x$, or, upon adding x to both sides, $7=13$.

It may well be argued, 'But how do we *know* that $4x-40$, the numerator on both sides of (3), is equal to zero?' This question brings up another point which was briefly noted at the beginning of the present chapter. There it was pointed out that the axioms cannot be applied blindly to equations without taking into consideration the values of the variables for which the equations are true. Thus equation (1), unlike the initial equations in Paradox 2 above, is not an identity which is true for all values of x, but is an equation which is satisfied only for the value $x = 10$. To verify this statement, clear of fractions in (3), getting successively

$$(13-x)(4x-40) = (7-x)(4x-40),$$
$$(4x-40)(13-x-7+x) = 0,$$
$$24(x-10) = 0,$$
$$x = 10.$$

Consequently the only value of x for which the equation is true is $x = 10$, and this reduces the numerators in (3) to zero.

In the following three problems[7] we shall have occasion to use certain properties of proportions. We recall them now for our convenience. If $\dfrac{p}{q} = \dfrac{r}{s}$, then it follows that

$$(A) \quad \frac{p-q}{q} = \frac{r-s}{s},$$

$$(B) \quad \frac{p}{q-p} = \frac{r}{s-r},$$

$$(C) \quad \frac{p-r}{q-s} = \frac{p}{q} = \frac{r}{s}.$$

To prove the first, note that if $\dfrac{p}{q} = \dfrac{r}{s}$, then, subtracting 1 from both sides, $\dfrac{p}{q} - 1 = \dfrac{r}{s} - 1$, whence $\dfrac{p-q}{q} = \dfrac{r-s}{s}$. The others can be proved similarly.

Paradox 1. Consider the proportion

$$\frac{x+1}{a+b+1} = \frac{x-1}{a+b-1}.$$

Applying property (A), we have

$$\frac{x+1-(a+b+1)}{a+b+1} = \frac{x-1-(a+b-1)}{a+b-1},$$

or

$$\frac{x-a-b}{a+b+1} = \frac{x-a-b}{a+b-1}.$$

89

And applying property (B),

$$\frac{x+1}{a+b+1-(x+1)}=\frac{x-1}{a+b-1-(x-1)},$$

or

$$\frac{x+1}{a+b-x}=\frac{x-1}{a+b-x}.$$

In the first result the numerators are equal. So, then, are the denominators. Hence $a+b+1=a+b-1$, or $+1=-1$. In the second result the denominators are equal. So, then, are the numerators. Hence $x+1=x-1$, and again $+1=-1$.

Paradox 2. Suppose that

$$\frac{3x-b}{3x-5b}=\frac{3a-4b}{3a-8b}.$$

These fractions are obviously (?) different from unity. But if we apply property (C) we must conclude that

$$\frac{3x-b-(3a-4b)}{3x-5b-(3a-8b)}, \text{ or } \frac{3x-3a+3b}{3x-3a+3b}, \text{ or } 1,$$

is equal to each of the original fractions. That is to say,

$$\frac{3x-b}{3x-5b}=\frac{3a-4b}{3a-8b}=1.$$

Paradox 3. Consider the proportions

$$\frac{x-a+c}{y-a+b}=\frac{b}{c} \text{ and } \frac{x+c}{y+b}=\frac{a+b}{a+c}.$$

Applying property (C) to each, we have

$$\frac{x-a-b+c}{y-a+b-c}=\frac{b}{c} \text{ and } \frac{x-a-b+c}{y-a+b-c}=\frac{a+b}{a+c}.$$

Hence b/c and $(a+b)/(a+c)$ are each equal to a third fraction without being equal to each other.

*

Before we leave the subject of equations which are not identities, let us look at an example or two of the ways in which hidden contradictions in the equations can bring about contradictions in their solution. We had the very obvious case at the beginning of this chapter of the system of two simultaneous equations in two unknowns, $x+y=1$, $x+y=2$. It was pointed out at the time that

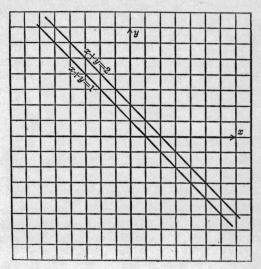

FIG. 61. The graphs of $x+y=1$ and $x+y=2$ are parallel straight lines and so have no points in common

there are no values of x and y for which $x+y$ is equal to both 1 and 2 at the same time. A graphical interpretation of this statement is given in Figure 61. But here is an example which is not quite so obvious.

$$\begin{cases} 2x+y=8; & (1) \\ x=2-\dfrac{y}{2}. & (2) \end{cases}$$

Substitute (2) in (1). It follows that $4-y+y=8$, or $4=8$. To discover the trouble here it is only necessary to clear of fractions in (2) and to add y to both sides. The system is then seen to be

$$\begin{cases} 2x+y=8; & (1) \\ 2x+y=4. & (2) \end{cases}$$

Paradox. The following system of equations[8] is of what is known as the 'homogeneous' type. We follow the usual method of solving such a system.

$$\begin{cases} 2x^2 - 3xy + y^2 = 4; & (1) \\ x^2 + 2xy - 3y^2 = 9. & (2) \end{cases}$$

Multiply both sides of (1) and (2) by 9 and 4 respectively. Since the resulting right-hand sides are equal, so also are the resulting left-hand sides. That is to say,

$$9(2x^2 - 3xy + y^2) = 4(x^2 + 2xy - 3y^2).$$

Simplifying,

$$2x^2 - 5xy + 3y^2 = 0.$$

Factorizing,

$$(2x - 3y)(x - y) = 0.$$

Now the product of two factors will be zero if either of the factors is zero. Hence

$$2x - 3y = 0 \text{ or } x - y = 0.$$

Each of these equations is to be solved simultaneously with either (1) or (2). Substituting $y = 2x/3$ in either, we obtain the correct solutions $x = 3$, $y = 2$, and $x = -3$, $y = -2$. But if we substitute $y = x$ in (1), we get $0 = 4$; in (2), $0 = 9$.

*

Using an argument which involved division by zero, we have already proved that any two unequal numbers are equal to each other (Paradox 1, page 86). Here is a different proof of the same proposition.[9]

Let a and b be two unequal numbers, and let c be their arithmetic mean, or average (for example, if $a = 2$ and $b = 4$, then $c = [a+b]/2 = 3$). Then

$$\frac{a+b}{2} = c, \text{ or } a + b = 2c. \qquad (1)$$

Multiply both sides by $a - b$:

$$a^2 - b^2 = 2ac - 2bc. \qquad (2)$$

Add $b^2 - 2ac + c^2$ to both sides:

$$a^2 - 2ac + c^2 = b^2 - 2bc + c^2. \qquad (3)$$

Both sides of (3) are now perfect squares and can be written in the form

$$(a-c)^2 = (b-c)^2. \tag{4}$$

Take the square root of both sides. Then

$$a-c = b-c, \tag{5}$$

or

$$a = b. \tag{6}$$

We started with the assumption that a was *not* equal to b and have come to the conclusion that a *is* equal to b.

The difficulty is again one which was mentioned briefly at the beginning of this chapter. That is, in the extraction of a square root both signs must be taken into consideration, and the one which leads to a contradictory result such as ours must be rejected. In passing from step (4) to step (5), only the positive signs were used. Had we written (5) as $a-c = -(b-c)$, we should have obtained our original expression, $a+b = 2c$. The whole argument was purposely made somewhat involved. It could have been done more obviously in this manner:

$$a+b = 2c,$$
$$a-c = c-b,$$
$$(a-c)^2 = (c-b)^2,$$
$$= (b-c)^2,$$
$$a-c = b-c,$$
$$a = b.$$

Paradox. To Prove that $n = n+1$.[10]

For any value of n, the identity

$$(n+1)^2 = n^2 + 2n + 1 \tag{1}$$

is true. Subtracting $2n+1$ from both sides,

$$(n+1)^2 - (2n+1) = n^2. \tag{2}$$

Subtracting $n(2n+1)$ from both sides,

$$(n+1)^2 - (2n+1) - n(2n+1) = n^2 - n(2n+1). \tag{3}$$

Adding $(2n+1)^2/4$ to both sides,

$$(n+1)^2 - (n+1)(2n+1) + \frac{(2n+1)^2}{4} = n^2 - n(2n+1) + \frac{(2n+1)^2}{4}. \tag{4}$$

Both members are now perfect squares, and can be written

$$\left[(n+1)-\left(\frac{2n+1}{2}\right)\right]^2 = \left[n-\left(\frac{2n+1}{2}\right)\right]^2. \tag{5}$$

Extracting the square root of both sides,

$$n+1-\left(\frac{2n+1}{2}\right) = n-\left(\frac{2n+1}{2}\right), \tag{6}$$

or, adding $(2n+1)/2$ to both sides,

$$n+1 = n.$$

*

If we think hard enough, we shall perhaps recall that in addition to the axioms concerning equalities, we once had to memorize a number of axioms concerning inequalities. They went something like this, did they not: 'Unequals plus or minus equals are unequal in the same order; unequals multiplied or divided by equals are unequal in the same order'; and so on? Did someone say that the divisors must not be zero? Good. But does anyone remember any other condition we put on that same axiom? Well, it's going to turn up in just a moment. But first let us recall the symbols used. '$a > b$' means 'a is greater than b'; '$a < b$' means 'a is less than b.'[11]

Now, assume that n and a are both positive integers. Then certainly

$$2n-1 < 2n. \tag{1}$$

Multiply both sides by $-a$. Then

$$-2an+a < -2an. \tag{2}$$

Add $2an$ to both sides:

$$+a < 0. \tag{3}$$

But this means that a, which we specified was positive, is negative. Now does anyone remember that additional condition? It is to the effect that the quantities by which we multiply or divide both sides of an inequality shall be *positive*, and we multplied by a negative number in step (2). Try the next two problems yourselves.

Paradox 1. To prove that any number is greater than itself. Assume that a and b are positive, and that

$$a > b. \tag{1}$$

94

Mutliplying both sides by b,

$$ab > b^2. \tag{2}$$

Subtracting a^2 from both sides, and factorizing,

$$a(b-a) > (b+a)(b-a). \tag{3}$$

Dividing both sides by $b-a$,

$$a > b+a. \tag{4}$$

Then since b is positive, not only is a greater than itself, but greater than any number greater than itself!

Paradox 2. To prove that $\frac{1}{8} > \frac{1}{4}$.
We must make use here of the following property of logarithms: $n \log (m) = \log (m)^n$. We start with the inequality

$$3 > 2. \tag{1}$$

Multiply both sides by $\log (\frac{1}{2})$. Then

$$3 . \log (\tfrac{1}{2}) > 2 . \log (\tfrac{1}{2}), \tag{2}$$

or

$$\log (\tfrac{1}{2})^3 > \log (\tfrac{1}{2})^2. \tag{3}$$

Whence

$$(\tfrac{1}{2})^3 > (\tfrac{1}{2})^2, \text{ or } \tfrac{1}{8} > \tfrac{1}{4}.$$

*

A number of fallacious results arise in connection with imaginary numbers – that is to say, square roots of negative numbers. The term 'imaginary' is unfortunate, but it is a term which has stuck with such numbers since they were first introduced. Until the beginning of the seventeenth century mathematicians worked for the most part with positive numbers only. Negative numbers were called 'absurd' and 'fictitious', and imaginary numbers were generally rejected as impossible. Actually, the number $\sqrt{-1}$ is no more imaginary than the number -1, which in turn is no more imaginary than the number 1. The concept of number is a complex one, [12] and we have no time to follow its complexities here, although we shall do so, in a small way, in Chapter 7. But as far as practicability is concerned, imaginary numbers have been found to be indispensable in such things as the development of communication by radio, telegraph, and telephone, and in the development of modern electrical methods of prospecting for oil.

Imaginary numbers arose from the demand of the mathematician that the equation $x^2 = a$ should *always* have a solution. Thus, if $x^2 = 1$ has a solution, why not $x^2 = -1$? The square root of -1 is defined in the same way as the square root of any positive number. That is to say, $\sqrt{-1}$ is that number which, when squared, gives -1. (Compare with $\sqrt{4}$, or 2, which, when squared, gives 4.) The square root of any negative number such as $-a$ (where a is positive) can be written as the real number, $\sqrt{a}$, times $\sqrt{-1}$, and for convenience $\sqrt{-1}$ is usually denoted by i. Thus

$$\sqrt{-a} = \sqrt{-1} \cdot \sqrt{a} = i\sqrt{a}.$$

Here we shall, for simplicity's sake, restrict our attention to positive square roots.

A contradiction which every student runs into, when he is first introduced to imaginaries, occurs when he attempts to apply to them the usual rules for the multiplication of radicals. He has learned that $\sqrt{a} \cdot \sqrt{b} = \sqrt{ab}$. For example, $\sqrt{2} \cdot \sqrt{3} = \sqrt{6}$. But this gives

$$\sqrt{-1} \cdot \sqrt{-1} = \sqrt{(-1)(-1)} = \sqrt{1} = 1;$$

whereas, by definition, $\sqrt{-1} \cdot \sqrt{-1} = -1$. Hence $-1 = +1$. The only way out of this difficulty is to agree *not* to apply the ordinary rules for radicals to imaginary numbers. The difficulty is generally avoided by writing i for $\sqrt{-1}$ and replacing i^2, wherever it appears, by -1, its true value by definition.

Paradox 1. A second proof that $-1 = +1$.[13]

We have, successively,

$$\sqrt{-1} = \sqrt{-1}, \tag{1}$$

$$\sqrt{\frac{1}{-1}} = \sqrt{\frac{-1}{1}}, \tag{2}$$

$$\frac{\sqrt{1}}{\sqrt{-1}} = \frac{\sqrt{-1}}{\sqrt{1}}, \tag{3}$$

$$\sqrt{1} \cdot \sqrt{1} = \sqrt{-1} \cdot \sqrt{-1}, \tag{4}$$

$$1 = -1. \tag{5}$$

Paradox 2. A third proof that $-1 = +1$.[14]

Consider the following, which is an identity for all values of x and y:

$$\sqrt{x-y} = i\sqrt{y-x}. \qquad \cdot \quad (1)$$

Substituting $x=a$, $y=b$,

$$\sqrt{a-b} = i\sqrt{b-a}. \qquad (2)$$

Substituting $x=b$, $y=a$,

$$\sqrt{b-a} = i\sqrt{a-b}. \qquad (3)$$

Mutliplying (2) by (3),

$$\sqrt{a-b} \cdot \sqrt{b-a} = i^2 \cdot \sqrt{b-a} \cdot \sqrt{a-b}. \qquad (4)$$

Dividing both sides by $\sqrt{a-b}$ and $\sqrt{b-a}$,

$$1 = i^2, \qquad (5)$$

or

$$1 = -1.$$

Geometrical Fallacies

THE fallacies of geometry are more remarkable than those of algebra in at least one respect, for the deception is not only of the mind, but of the eye as well. The diagrams in Figure 18 at the beginning of Chapter 4 showed how easy it is for the eye to mislead the mind. The examples in the last chapter furnished scant material for the eye, but did reveal that care must be used if the mind is not to lead itself astray. Formal deduction in geometry is to some extent a combination of seeing and reasoning, for in the proof of any theorem the logical processes of the mind are guided by and checked against what the eye sees in the figure.

It may be of interest to note that Euclid compiled a collection of exercises for the detection of fallacies, but unfortunately this work has been lost.[1]

*

As our first example of one type of fallacious geometrical reasoning we shall carry through a complete discussion of this remarkable theorem:

To prove that any triangle is isosceles.[2]

Let *ABC* be any triangle, as in Figure 62(a). Construct the bisector of $\angle C$ and the perpendicular bisector of side *AB*. From *G*, their point of intersection, drop perpendiculars *GD* and *GF* to

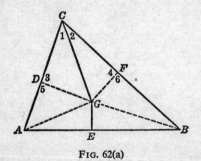

FIG. 62(a)

AC and *BC* respectively and draw *AG* and *BG*. Now in triangles *CGD* and *CGF*, ∠1 = ∠2 by construction and ∠3 = ∠4 since all right angles are equal. Furthermore the side *CG* is common to the two triangles. Therefore triangles *CGD* and *CGF* are congruent – can be made to coincide, that is. (If two angles and a side of one triangle are equal respectively to two angles and a side of another, the triangles are congruent.) It follows that *DG = GF*. (Corresponding parts of congruent triangles are equal.) Then in triangles *GDA* and *GFB*, ∠5 and ∠6 are right angles and, since *G* lies on the perpendicular bisector of *AB*, *AG = GB*. (Any point on the perpendicular bisector of a line is equidistant from the ends of the line.) Therefore triangles *GDA* and *GFB* are congruent. (If the hypotenuse and another side of one right triangle are equal respectively to the hypotenuse and another side of a second, the triangles are congruent.) From these two sets of congruent triangles – *CGD* and *CGF*, and *GDA* and *GFB* – we have, respectively,

$$CD = CF \qquad (1)$$

and

$$DA = FB. \qquad (2)$$

Adding (1) and (2), we conclude that *CA = CB*, so that triangle *ABC* is isosceles by definition.

It may be argued that we do not know that *EG* and *CG* meet within the triangle. Very well, then, we shall examine all other possibilities. The above proof, word for word, is valid in the cases wherein *G* coincides with *E*, or *G* is outside the triangle but so near to *AB* that *D* and *F* fall on *CA* and *CB* and not on *CA* and *CB* produced. These cases are illustrated in Figures 62(b) and (c).

There remains the possibility, shown in Figure 62(d), in which

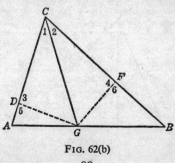

Fig. 62(b)

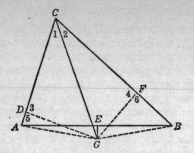

Fig. 62(c)

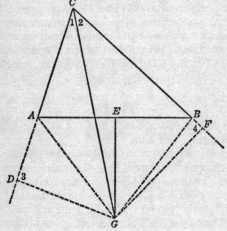

Fig. 62(d)

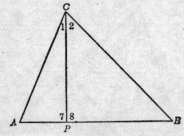

Fig. 62(e)

100

G lies so far outside the triangle that *D* and *F* fall on *CA* and *CB* produced. Again, as in the first case, triangles *CGD* and *CGF* are congruent, as are triangles *GDA* and *GFB*. And again *CD=CF* and *DA=FB*. But in the present case we must subtract these last two equations in order to have *CA=CB*.

Finally, it may be suggested that *CG* and *EG* do not meet in a single point *G*, but either coincide or are parallel. A glance at Figure 62(e) shows that in either of these cases the bisector *CP* of angle *C* will be perpendicular to *AB*, so that ∠7 = ∠8. Then ∠1 = ∠2, *CP* is common, and triangle *APC* is congruent to triangle *BPC*. Again *CA=CB*.

It certainly *appears* that we have exhausted all possibilities and that we must accept the obviously absurd conclusion that all triangles are isosceles. There is one more case, however, which may be worth investigating. Is it not possible for *one* of the points *D* and *F* to fall *inside* the triangle and for the *other* to fall *outside*? A correctly drawn figure will indicate that this possibility is indeed the only one. Furthermore we can prove it as follows.

Circumscribe a circle about the triangle *ABC*, as in Figure 62(f). Since ∠1 = ∠2, *CG* must bisect arc *AB*. (∠1 and ∠2 are inscribed

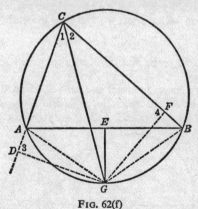

FIG. 62(f)

angles and, being equal, must be subtended by equal arcs.) But *EG* also bisects arc *AB*. (The perpendicular bisector of a chord bisects the arc of the chord.) It follows that *G* lies on the circumscribed circle and that *CAGB* is an inscribed quadrilateral. Now

$\angle CAG + \angle CBG$ is a straight angle. (The opposite angles of an inscribed quadrilateral are supplementary.) But if $\angle CAG$ and $\angle CBG$ were both right angles, D and F would coincide with A and B respectively; so the conclusion that $CD = CF$ (a conclusion established in the first case) would reduce to $CA = CB$, which is contrary to our hypothesis that ABC is *any* triangle. Consequently one of the angles CAG and CBG must be acute and the other obtuse, which means that either D or F (D in the figure) must fall outside the triangle and the other inside. The relations $CD = CF$ and $DA = FB$ are true here, as they were in all of our other cases. But whereas $CB = CF + FB$, we now have $CA = CD - DA$, not $CD + DA$.

This discussion has been lengthy, but it should have been instructive. It shows how easily a logical argument can be swayed by what the eye sees in the figure and so emphasizes the importance of drawing a figure correctly, noting with care the relative positions of points essential to the proof. Had we at the start actually constructed – by means of ruler and compasses – the angle bisector and the various perpendiculars, we should have saved ourselves a good deal of trouble.

The following five fallacies are all concerned with the same pitfalls as the one we have just worked over in detail. Watch your step!

Paradox 1. To prove that there are two perpendiculars from a point to a line.[3]

Let any two circles intersect in Q and R. Draw diameters QP and QS and let PS cut the circles at M and N respectively, as in Figure 63. Then $\angle PNQ$ and $\angle SMQ$ are right angles. (An angle

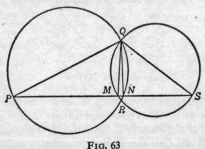

F IG . 63

inscribed in a semicircle is a right angle.) Hence *QM* and *QN* are both perpendicular to *PS*.

Paradox 2. To prove that a right angle is equal to an obtuse angle.[4]

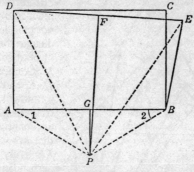

Fig. 64

Let *ABCD* be any rectangle. Following Figure 64, draw through *B* a line *BE* outside the rectangle and equal in length to *BC* (hence to *AD*). Construct the perpendicular bisectors of *DE* and *AB*. Since these lines are perpendicular to non-parallel lines, they must meet, as at *P*. Draw *AP*, *BP*, *DP*, and *EP*. In triangles *APD* and *BPE*, *AD*=*BE* by construction. Also *AP*=*BP* and *DP*=*EP*. (Any point in the perpendicular bisector of a line is equidistant from the ends of the line.) Since the three sides of triangle *APD* are equal respectively to the three sides of triangle *BPE*, these triangles are congruent. Hence

$$\angle DAP = \angle EBP. \tag{1}$$

But

$$\angle 1 = \angle 2. \tag{2}$$

(Angles opposite the equal sides of an isosceles triangle are equal.) Subtracting (2) from (1), we conclude that ∠*DAG* (given a right angle) is equal to ∠*EBG* (an obtuse angle by construction).

Paradox 3. To prove that 45° = 60°, or that 3=4.[5]

On side *AB* of equilateral triangle *ABC* as hypotenuse construct an isosceles right triangle *ABD*. We shall prove that ∠*ABC*, which is 60°, is equal to ∠*ABD*, which is 45°.

On BC lay off BE equal to BD. Mark F, the mid-point of AD, and through E and F draw a line which intersects BA produced in

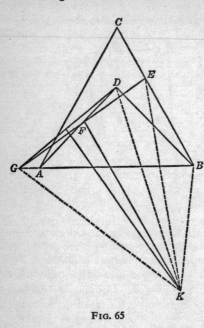

G. Draw GD. Then construct the perpendicular bisectors of GE and GD. Since GE and GD are not parallel, the perpendicular bisectors must meet at some point, say K. Connect K with G, D, E, and B. Our main job now is to show that triangles KDB and KEB are congruent. Note first that $KG=KD$ and $KG=KE$. (Any point in the perpendicular bisector of a line is equidistant from the ends of the line.) Hence $KD=KE$. Furthermore, $BD=BE$ by construction, and BK is a common side. Therefore triangle KDB is congruent to triangle KEB, whence $\angle KBD=$

FIG. 65

$\angle KBE$. If now we subtract the common portion $\angle KBA$ from each of these angles, we must conclude that $\angle ABD=\angle ABC$, or that $45° = 60°$, or that $3 = 4$.

Paradox 4. To prove that if two opposite sides of a quadrilateral are equal, the remaining two sides must be parallel.[6]

Suppose the quadrilateral is $ABCD$, as shown in Figure 66(a), with $AD=BC$. We shall prove that AB is parallel to DC. Erect the perpendicular bisectors of AB and DC. (In the figure, P and Q are the mid-points of DC and AB respectively.) If the perpendicular bisectors coincide or are parallel, then AB and DC, being perpendicular to the same line or to parallel lines, will be parallel, and the theorem is proved. So let us suppose that they meet at O. Draw OD, OC, OA, and OB.

Now triangles *DPO* and *CPO* are congruent, since *PO* is common and, by construction, $DP=PC$ and $\angle DPO = \angle CPO$. (If two sides and the included angle of one triangle are equal respectively to two sides and the included angle or another, the triangles are congruent.) Therefore $DO=CO$. In precisely the same manner triangles *AQO* and *BQO* are congruent and $AO=OB$. Also *AD* was given equal to *BC*. Therefore triangles *AOD* and *BOC* are congruent. (If the three sides of one triangle are equal respectively

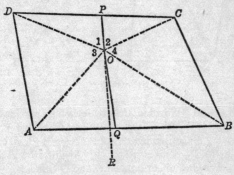

FIG. 66(a)

to the three sides of another, the triangles are congruent.) It follows from the congruence of triangles *DPO* and *CPO* that $\angle 1 = \angle 2$, and from the congruence of triangles *AOD* and *BOC* that $\angle 3 = \angle 4$. Hence $\angle 1 + \angle 3 = \angle 2 + \angle 4$. But if *OR* is the extension of *PO*, then $\angle 1 + \angle 3 + \angle AOR = \angle 2 + \angle 4 + \angle BOR$ because each of these sums is equivalent to a straight angle. Subtracting the first of these last two equations from the second, $\angle AOR = \angle BOR$. That is, *PO* extended bisects $\angle AOB$. On the other hand, from the congruence of triangles *AQO* and *BQO* it is evident that *OQ* bisects $\angle AOB$. Therefore *PR* and *OQ* must coincide, in which case *AB* and *DC* are both perpendicular to the same straight line and so must be parallel.

If *O* lies outside the quadrilateral, as in Figure 66(b), we have $\angle 1 = \angle 2$, $\angle 3 = \angle 4$ precisely as before. But now $\angle 1 - \angle 3 = \angle 2 - \angle 4$, or $\angle AOP = \angle BOP$. That is, *OP* again bisects $\angle AOB$. But, as before, so also does *OQ*. Therefore *OP* and *OQ* coincide and again *AB* is parallel to *DC*.

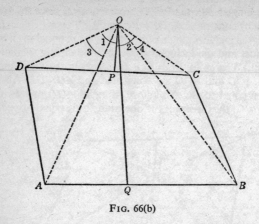

FIG. 66(b)

Finally, if O and P coincide, then by reasoning similar to what has gone before it is easy to show in Figure 66(c) that $\angle 3 = \angle 4$ and that $\angle 5 = \angle 6$. Consequently $\angle 3 + \angle 5 = \angle 4 + \angle 6$, or OQ is perpendicular to DC as well as to AB. Again AB and DC are parallel. A similar argument holds if O and Q coincide.

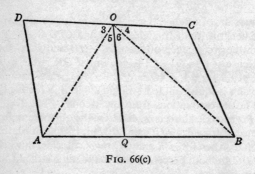

FIG. 66(c)

Paradox 5. To prove that every point inside a circle must lie on the circumference of the circle.[7]

Let B be any point within the circle O of Figure 67. Through B draw diameter AC. Now find point D on AC produced so that D divides AC externally in the same ratio that B divides AC internally

106

– in other words, so that $AB/BC=AD/DC$. Erect QP, the perpendicular bisector of BD, and draw OP and BP.

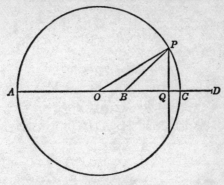

If r denotes the radius of the circle, then $AB=r+OB$, $BC=r-OB$, $AD=OD+r$, and $DC=OD-r$. The proportion $AB/BC=AD/DC$ can then be written

$$\frac{r+OB}{r-OB}=\frac{OD+r}{OD-r}. \tag{1}$$

Clearing of fractions

$$(r+OB)(OD-r)=(r-OB)(OD+r). \tag{2}$$

Multiplying out and simplifying,

$$OB.OD=r^2. \tag{3}$$

Now from the figure,

$$OB=OQ-BQ, \tag{4}$$

and

$$OD=OQ+QD. \tag{5}$$

But since Q bisects BD, QD is equal to BQ, so that (5) can be written in the form

$$OD=OQ+BQ. \tag{6}$$

Multiplying (4) by (6) and substituting r^2 for $OB.OD$ on the left,

$$r^2=\overline{OQ}^2-\overline{BQ}^2. \tag{7}$$

Applying the Pythagorean theorem (the square of the hypotenuse of a right triangle is equal to the sum of the squares of the other two sides) to triangles OQP and BQP, we have

$$\overline{OP}^2 = \overline{OQ}^2 + \overline{QP}^2, \tag{8}$$

$$\overline{BP}^2 = \overline{BQ}^2 + \overline{QP}^2. \tag{9}$$

Subtracting (9) from (8),

$$\overline{OP}^2 - \overline{BP}^2 = \overline{OQ}^2 - \overline{BQ}^2. \tag{10}$$

But $OP = r$. Therefore (10) can be written

$$r^2 - \overline{BP}^2 = \overline{OQ}^2 - \overline{BQ}^2. \tag{11}$$

Now replace the right-hand side of (11) by its value given in (7). Then

$$r^2 - \overline{BP}^2 = r^2,$$

or

$$\overline{BP}^2 = 0,$$

whence

$$BP = 0.$$

But if $BP = 0$, it follows that B and P must coincide, so that B, given as any *interior* point of the circle, must lie on the circumference of the circle.

*

The following group of problems is concerned with a type of fallacy which we discussed at length in Chapter 5. We shall perhaps recognize the error when it turns up. Consider the following theorem.

To prove that two unequal lines are equal.[8]

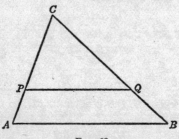

FIG. 68

Let ABC be any triangle and draw any line PQ parallel to AB. Then triangles ABC and PQC are similar. (If a line is drawn parallel to one side of a triangle and intersecting the other two sides, it cuts off a triangle similar to the given one.) Consequently

$$\frac{AB}{PQ}=\frac{AC}{PC}. \tag{1}$$

(The corresponding sides of two similar triangles are proportional by definition.) That is,

$$AB.PC=AC.PQ. \tag{2}$$

Multiply both sides by $AB-PQ$:

$$\overline{AB}^2.PC-AB.PC.PQ=AB.AC.PQ-\overline{PQ}^2.AC. \tag{3}$$

Add $AB.PC.PQ$ to both sides and subtract $AB.AC.PQ$ from both sides:

$$\overline{AB}^2.PC-AB.AC.PQ=AB.PC.PQ-\overline{PQ}^2.AC. \tag{4}$$

Factorize:

$$AB(AB.PC-AC.PQ)=PQ(AB.PC-AC.PQ). \tag{5}$$

Divide both sides by $AB.PC-AC.PQ$. Then

$$AB=PQ. \tag{6}$$

This proof is perhaps rather convincing. The figure is so simple that no error can lie in that direction, and the logical argument is straightforward. But in fact, it's our old friend, division by zero. In step (2) we established the fact that $AB.PC=AC.PQ$, and in step (6) we divided both sides of the equation by the difference of these two equal quantities.

Some readers may have found this example a little obvious. What about the next two?

Paradox 1. To prove that a line segment is equal to part of itself.[9]

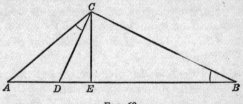

Fig. 69

In any triangle ABC, suppose that angle A is acute and that angle C is greater than angle B. (These suppositions in no way restrict the choice of a triangle, but merely constitute directions for lettering Figure 69.) Construct $\angle ACD$ equal to $\angle B$ and draw CE perpendicular to AB. We shall prove that $AB=AD$.

In triangles ABC and ADC, $\angle A$ is common, and $\angle B = \angle ACD$ by construction. The triangles are therefore similar. (If two angles of one triangle are equal respectively to two angles of another, the triangles are similar.) It follows that

$$\frac{\triangle ABC}{\triangle ADC} = \frac{\overline{CB}^2}{\overline{CD}^2}. \tag{1}$$

(If two triangles are similar, their areas are to each other as the squares of any two corresponding sides.) Moreover, since CE is the common altitude of the two triangles,

$$\frac{\triangle ABC}{\triangle ADC} = \frac{AB}{AD}. \tag{2}$$

(The areas of two triangles with equal altitudes are to each other as their bases.) From (1) and (2) it follows that

$$\frac{\overline{CB}^2}{\overline{CD}^2} = \frac{AB}{AD}. \tag{3}$$

Multiplying both sides of (3) by CD^2 and dividing both sides by AB,

$$\frac{\overline{CB}^2}{AB} = \frac{\overline{CD}^2}{AD}. \tag{4}$$

Now a theorem which is not included in all elementary texts on plane geometry is the following: *In any triangle the square of the side opposite an acute angle is equal to the sum of the squares of the other two sides minus twice the product of one of these sides times the projection of the other upon it.* (In the figure AE is the projection of AC on AB. Incidentally, this theorem is the basis of the law of cosines in trigonometry.) Applying this theorem to each of the triangles ABC and ADC, we can substitute for $\overline{CB}^2$ and $\overline{CD}^2$ in (4) as follows:

$$\frac{\overline{AC}^2 + \overline{AB}^2 - 2AB.AE}{AB} = \frac{\overline{AC}^2 + \overline{AD}^2 - 2AD.AE}{AD}. \tag{5}$$

Carrying out the indicated division, we can write (5) in the form

$$\frac{\overline{AC}^2}{AB} + AB - 2AE = \frac{\overline{AC}^2}{AD} + AD - 2AE. \tag{6}$$

Adding 2*AE* to both sides and subtracting *AB* and *AD* from both sides,

$$\frac{\overline{AC}^2}{AB} - AD = \frac{\overline{AC}^2}{AD} - AB, \tag{7}$$

or

$$\frac{\overline{AC}^2 - AB.AD}{AB} = \frac{\overline{AC}^2 - AB.AD}{AD}. \tag{8}$$

Since the numerators of the fractions in (8) are equal, so also are the denominators. That is, *AB*=*AD*.

Paradox 2. To prove that the sum of the two parallel sides of a trapezium is zero.[10]

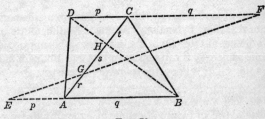

FIG. 70

Denote the parallel sides of trapezium *ABCD* by *p* and *q* as shown in Figure 70. Produce *DC* a distance *q* to *F*, and *BA* a distance *p* to *E*. Draw *EF*, *DB*, and *AC*; and denote *AG*, *GH*, and *HC*, the segments into which *AC* is divided, by *r*, *s*, and *t* respectively.

In triangles *ABH* and *CDH*, ∠*HAB*=∠*HCD* and ∠*HBA* =∠*HDC*. (If two parallel lines are cut by a transversal, the alternate interior angles are equal.) Hence triangles *ABH* and *CDH* are similar. (If two angles of one triangle are equal respectively to two angles of another, the triangles are similar.) Therefore

$$\frac{DC}{AB} = \frac{HC}{HA}, \text{ or } \frac{p}{q} = \frac{t}{r+s}. \tag{1}$$

(By definition, the corresponding sides of two similar triangles are proportional.) In exactly the same way it can be shown that triangle *EAG* is similar to triangle *FCG* and that

$$\frac{AE}{CF} = \frac{AG}{GC}, \text{ or } \frac{p}{q} = \frac{r}{s+t}. \tag{2}$$

From (1) and (2) it follows that

$$\frac{p}{q} = \frac{t}{r+s} = \frac{r}{s+t}. \tag{3}$$

Now apply to the second and third members of (3) one of the properties of proportions listed in Chapter 5. (Property C, page 89.) This operation gives

$$\frac{p}{q} = \frac{t-r}{r+s-(s+t)} = \frac{t-r}{r-t} = \frac{-(r-t)}{r-t} = -1. \tag{4}$$

From (4) we conclude that $p = -q$, or that $p+q = 0$. In other words, the sum of the sides DC and AB of trapezium $ABCD$ is zero.

*

A great deal of unnecessary writing is avoided in mathematics by the use of reasoning by analogy. We made legitimate use of this type of argument in step (2) of the last example when we said, 'In exactly the same way it can be shown that . . .' But care must be used in applying it. Witness the following theorem:

To prove that $\sqrt{a} + \sqrt{b} = \sqrt{2(a+b)}$.[11]

In triangle ABC of Figure 71 denote by h the altitude from C to AB, and by p and q the segments into which AB is divided by the

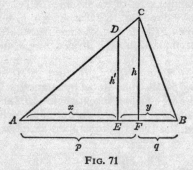

FIG. 71

altitude. Now construct a line h' so that it will be parallel to h and will divide the triangle into two parts of equal area. Call the segments into which AB is divided by this line x and y respectively. Then

$$2. \triangle AED = \triangle ABC. \tag{1}$$

Since the area of a triangle is equal to one half the base times the altitude, (1) can be written

$$2 \cdot \tfrac{1}{2}xh' = \tfrac{1}{2}(p+q)h. \tag{2}$$

Now triangle AED is similar to triangle AFC. (If a line is drawn parallel to one side of a given triangle and intersecting the other two sides, it cuts off a triangle similar to the given one.) Consequently

$$\frac{h'}{h} = \frac{x}{p}. \tag{3}$$

(By definition, the corresponding sides of two similar triangles are proportional.) Solving (3) for h' and substituting in (2),

$$\frac{x^2 h}{p} = \tfrac{1}{2}(p+q)h. \tag{4}$$

Dividing both sides of (4) by h, multiplying both sides by p, and extracting the square root of both sides,

$$x = \sqrt{\frac{p(p+q)}{2}}. \tag{5}$$

(Here we are justified in taking only the positive sign with the square root, since x is a segment of positive length.) Now y bears the same relation to q that x bears to p. By similar reasoning, then, we have

$$y = \sqrt{\frac{q(p+q)}{2}}. \tag{6}$$

Adding (5) and (6) and replacing $x+y$ on the left by $p+q$ (each of these quantities is identical with AB, the base of the original triangle),

$$p+q = \sqrt{\frac{p(p+q)}{2}} + \sqrt{\frac{q(p+q)}{2}} \tag{7}$$

$$= \sqrt{p+q}\left(\sqrt{\frac{p}{2}} + \sqrt{\frac{q}{2}}\right). \tag{8}$$

Dividing both sides of (8) by $\sqrt{p+q}$,

$$\sqrt{p+q} = \sqrt{\frac{p}{2}} + \sqrt{\frac{q}{2}}. \tag{9}$$

Finally, substituting $2a$ for p and $2b$ for q,

$$\sqrt{2(a+b)} = \sqrt{a} + \sqrt{b}.$$

This result, of course, is ridiculous. The error occurred in step (6). We cannot reason about y as we did about x. We made use in (2) and (3) of the fact that x is the base of a triangle similar to triangle AFC, and y does not enjoy this property. In other words, y does *not* bear the same relation to q that x bears to p.

Reasoning by analogy is safe enough if properly used, but even outside the field of mathematics it can lead, if misused, to results which are not only absurd, but sometimes disastrous.

*

We conclude this chapter with two fallacies for those of us who have studied solid geometry.

Paradox 1. To prove that the sum of the angles of a spherical triangle is 180°.[12]

Let ABC be any spherical triangle. Choose any point P inside the triangle and pass great circles through P and A, B, and C respectively, dividing the original spherical triangle into three smaller ones. (See Figure 72.) Now call the sum of the angles of any spherical triangle $x°$. Then the sum of the angles of the three small triangles is $3x°$. Included in this sum is the sum of the angles about

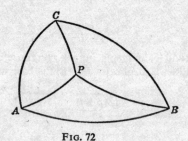

Fig. 72

the point P, or 360°. But the sum of the angles of triangle ABC is equal to the sum of the angles of the three small triangles minus the sum of the angles at P. That is, $x = 3x - 360$, whence $2x = 360$, or $x = 180$. This conclusion contradicts a well-known theorem to the effect that the sum of the angles of a spherical triangle can be anything between 180° and 540°.

Paradox 2. To prove that from a point outside a plane an infinite number of perpendiculars can be drawn to the plane.[13]

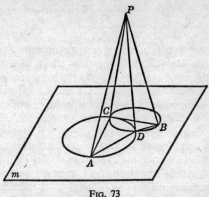

FIG. 73

In Figure 73 let P be any point outside of plane m. Choose any two points A and B in the plane, and on PA and PB as diameters construct two spheres. These spheres will intersect the plane m in two circles. (The intersection of a plane and a sphere is a circle.) And these two circles will intersect at two points, say C and D. Draw PC, PD, AC, AD, BC, and BD.

Now think of a plane passed through P, A, and C. (Three points determine a plane.) This plane will intersect the sphere about PA in a circle, so that $\angle PCA$ will be inscribed in a semicircle. Hence $\angle PCA$ is a right angle. (An angle inscribed in a semicircle is a right angle.) $\angle PCB$ is a right angle for the same reason. Therefore PC is perpendicular to both CA and CB, and so perpendicular to plane m. (If a line is perpendicular to each of two intersecting lines at their point of intersection, it is perpendicular to the plane of the two lines.) In exactly the same way it can be shown that PD is perpendicular to both DA and DB, and so perpendicular to plane m. But since there are an infinite number of choices for A and B, and since to each choice correspond two perpendiculars, there must be an infinite number of perpendiculars from P to m.

CHAPTER SEVEN

Paradoxes of the Infinite

FOR well over two thousand years mathematicians have been struggling with the infinite. They cannot afford to disregard it, for it is indispensable in much of their work. Yet in their attempts to understand it and to use it, they have run up against many contradictions. Some of these they have been able to overcome, while others are still causing them trouble. Indeed, the paradoxes enunciated by Zeno of Elea in the fifth century B.C. have never been settled to the complete satisfaction of all mathematicians.

The infinite is an insidious sort of monster. It often turns up when least expected – when one's back is turned, so to speak. Then, too, it is sometimes difficult to recognize, for there is more than one breed of the monster. There is the infinite in algebra, the infinite in geometry, the infinitely small, the infinitely large, and so on. Again, there is not only one infinite, but a whole hierarchy of infinites.

In a single chapter we cannot hope to cover material which has filled entire volumes. We shall go into the subject just far enough to be able to appreciate some of its remarkable paradoxes. Whenever possible, the reader will be referred to more detailed treatments of the various topics we discuss here.

THE INFINITE IN ARITHMETIC

First let us consider what we shall mean by an infinite class, or group, or collection of things. For present purposes the following intuitive and rather loose definition will do. 'An infinite class is one whose members cannot be counted in any finite period of time, however long.' Incidentally, we shall assume that the counting proceeds at a uniform rate – one member a second, let us say. Some of us will object to this definition on the ground that we use the finite to define the infinite, but we shall have to agree that everyone knows what 'a finite period of time' means.

We must not confuse the *infinite* with the *very large finite*. Think, for example, of the number of inhabitants of the earth at any particular instant, or the number of leaves on all the trees of the earth at any instant, or the number of blades of grass on the earth at any instant. These are all very large numbers, yet they are *finite*. That is to say, given sufficient patience and man-power, we could set out to count the members of these large classes with the assurance that we could finish the job. Some twenty-one centuries ago Archimedes showed that he was able to distinguish between the infinite and the large finite when he estimated the number of grains of sand required to fill the then known universe.

Where are we to find an example of an infinite class? Certainly not in our world of physical experience, which after all is a finite world. But wait. We have just spoken of counting the members of a large collection. What of the collection consisting of the very numbers with which we count – the so-called 'natural numbers'? Here is a class which fulfils the requirements of our definition. For if we set out to count the natural numbers, 1, 2, 3, 4, 5, . . ., can we not do so with the assurance that if we continue until we die, and pass the job on from generation to generation, neither we nor any of our descendants will ever exhaust the supply? The natural numbers, then, provide us with an infinite class with which we are fairly familiar.

Before going any further, let us examine a few other examples of infinite classes, noting that they all arise out of the fundamental natural numbers. In each case the dots signify, of course, that the sequence goes on indefinitely, that is, without end.

(1) All values of n^2, where n is a natural number:

$$1, 4, 9, 16, 25, 36, 49, 64, \ldots$$

(2) All values of $\frac{1}{n}$, where n is a natural number:

$$\frac{1}{1}, \frac{1}{2}, \frac{1}{3}, \frac{1}{4}, \frac{1}{5}, \frac{1}{6}, \frac{1}{7}, \frac{1}{8}, \ldots$$

(3) All values of 2^n, where n is a natural number:

$$2, 4, 8, 16, 32, 64, 128, 256, \ldots$$

(4) All values of $\frac{1}{2^n}$, where n is a natural number:

$$\frac{1}{2}, \frac{1}{4}, \frac{1}{8}, \frac{1}{16}, \frac{1}{32}, \frac{1}{64}, \frac{1}{128}, \frac{1}{256}, \ldots$$

All of these classes have the property that their members cannot be exhausted by counting over any finite period of time, however long.

*

We are now in a position to consider the first of the paradoxes of Zeno, mentioned briefly at the beginning of this chapter: *Motion is impossible*. The conclusion is startling, we must admit; and the argument is rather convincing. Let's look at it.

To go from any point P to another point Q, we must first go half the distance from P to Q, then half the remaining distance, then half the distance then remaining, then half the distance *then* remaining, and so on. The 'and so on' implies that the process can be repeated, and is to be repeated, an infinite number of times. Now regardless of how small the successive distances become, each one of them unquestionably requires a finite length of time to cover. And, argued Zeno, the sum of an infinite number of finite intervals of time must be infinite. Therefore we can never get from P to Q, however near together P and Q may be.

A number of possible solutions of this paradox have been proposed.[1] The one we shall choose attributes the fallacy to the statement 'the sum of an infinite number of finite intervals of time must be infinite.' This statement is generally, but not always, true. First let us investigate the sum of all the members of the infinite class in example (3) above. If we write

$$2+4+8+16+32+64+128+256+...,$$

it is evident at once that as we go on adding successive terms, the sum rapidly becomes larger and larger. Actually, it is not enough to say simply 'larger and larger'. We must be more precise. Let us note that by going out far enough in the series, we can make the sum of all the terms up to that point exceed any finite number, however large. This fact is indicated graphically in Figure 74. For example, if someone names the finite number 1000, we can, by taking 9 terms, make the sum 1022. If he raises the bid to 1,000,000 we can make the sum 1,048,574 by taking 19 terms. If he cares to go as high as 1,000,000,000, we have only to take 29 terms to make the sum 1,073,741,822. No matter *how* large a finite number our imaginary adversary sees fit to choose, it is evident that we can always make our sum exceed his number by taking a sufficiently large finite number of terms. This is what the mathematician means when he says that 'the sum of this infinite series is infinite.'

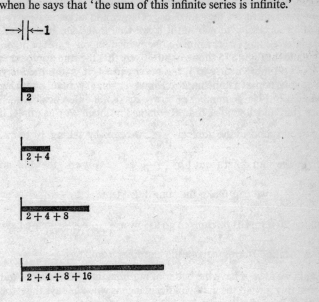

FIG. 74. The sum $2+4+8+16+32+64+128+256+\ldots$ increases without limit

But now let us return to the problem of motion from one point to another. Suppose the distance from P to Q is 100 yards, and that we walk at the rate of 100 yards per minute. Then the time required for the first stage of the journey – half the distance from P to Q – is $\frac{1}{2}$ minute; that for half the remaining distance, $\frac{1}{4}$ minute; that for half the distance then remaining, $\frac{1}{8}$ minute; that for half the distance *then* remaining, $\frac{1}{16}$ minute; and so on. In other words, the time in minutes required to go from P to Q is the sum of the infinite series

$$\frac{1}{2}+\frac{1}{4}+\frac{1}{8}+\frac{1}{16}+\frac{1}{32}+\frac{1}{64}+\frac{1}{128}+\frac{1}{256}+\ldots$$

(Note that this is the sum of all the members of the infinite class in example (4) above.) Is the sum of this infinite series infinite? As in our previous series, the sum does get larger and larger as we go on adding successive terms. But it is *not* true that we can make the sum exceed any larger finite number which anyone cares to name. A glance at Figure 75 shows us intuitively that the sum approaches more and more nearly to 1, but never exceeds it. More precisely, if anyone names a finite number, however small, we can, by taking a sufficiently large number of terms, make the difference between our sum and 1 smaller than the named number. For example, if someone chooses the number $\frac{1}{1000}$, we can, by taking 10 terms, make the sum differ from 1 by $\frac{1}{1024}$. If he lowers the bid to $\frac{1}{1,000,000}$, we can make the sum differ from 1 by $\frac{1}{1,048,576}$ by taking 20 terms. If he cares to go as low as $\frac{1}{1,000,000,000}$, we have only to take 30 terms to make the sum differ from 1 by $\frac{1}{1,073,741,824}$. Again we always have the better of our imaginary adversary. And again this is what the mathematician means when he says that 'the sum of this infinite series is 1.'

Consequently the time required to travel the 100 yards from *P* to *Q* is not infinite, but is 1 minute. Motion, we learn with some relief, is *not* impossible. Here mathematics comes to our aid and backs up what everyday experience has taught us.

*

Zeno's second paradox involves the problem of Achilles and the tortoise. The argument in this case is to the effect that if Achilles gives the tortoise a head start, he can never overtake him. For Achilles must always first get to the point from which the tortoise has just departed, and in this way the tortoise is always ahead.

To clarify our ideas, let us suppose that Achilles gives the tortoise a start of 100 yards, that Achilles travels at the rate of 10 yards per second, and that the tortoise travels at the rate of 1 yard per second. Then Achilles travels the first 100 yards in 10 seconds. In the meantime the tortoise has gone 10 yards. Achilles takes 1 second to cover that distance, while the tortoise advances 1 yard. Achilles covers

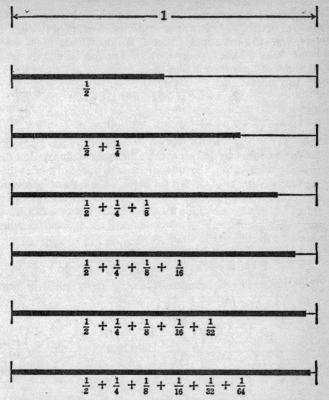

FIG. 75 The sum $\frac{1}{2}+\frac{1}{4}+\frac{1}{8}+\frac{1}{16}+\frac{1}{32}+\frac{1}{64}+\frac{1}{128}+\frac{1}{256}+\ldots$ approaches 1 as a limit

that distance in $\frac{1}{10}$ second, and the tortoise is still $\frac{1}{10}$ yard ahead. And so on. Then the number of seconds which elapse before Achilles catches up with the tortoise is the sum of the infinite series

$$10+1+\frac{1}{10}+\frac{1}{100}+\frac{1}{1000}+\ldots$$

For those of us who remember our formulas for geometric progressions, it is but a moment's labour to show that this sum is not infinite, but that it is $11\frac{1}{9}$ seconds.

*

121

Within the last hundred years numerous criteria have been developed to determine whether a given series 'diverges to infinity' or 'converges to a finite limit' – that is, whether the sum of the series is infinite or a finite number.[2] We shall not go far into the technicalities of these criteria, but let us return for just a moment to the two series,

$$2+4+8+16+32+64+128+...$$

and

$$\frac{1}{2}+\frac{1}{4}+\frac{1}{8}+\frac{1}{16}+\frac{1}{32}+\frac{1}{64}+\frac{1}{128}+...$$

We have seen that the first of these diverges to infinity, while the second converges to 1. Can this difference be traced to the fact that the successive terms of the first get *larger*, while those of the second get *smaller*? Let us not jump too hastily to conclusions. This much is true: a *necessary* condition for convergence is that the successive terms decrease in size. That this condition is *not sufficient* is readily shown by the 'harmonic series',

$$1+\frac{1}{2}+\frac{1}{3}+\frac{1}{4}+\frac{1}{5}+\frac{1}{6}+\frac{1}{7}+\frac{1}{8}+\frac{1}{9}+\frac{1}{10}+\frac{1}{11}+\frac{1}{12}+\frac{1}{13}+\frac{1}{14}+\frac{1}{15}+\frac{1}{16}+...$$

This series, can by the insertion of brackets, be written in the form

$$1+\frac{1}{2}+\left(\frac{1}{3}+\frac{1}{4}\right)+\left(\frac{1}{5}+\frac{1}{6}+\frac{1}{7}+\frac{1}{8}\right)+$$

$$\left(\frac{1}{9}+\frac{1}{10}+\frac{1}{11}+\frac{1}{12}+\frac{1}{13}+\frac{1}{14}+\frac{1}{15}+\frac{1}{16}\right)+...$$

Now since $\frac{1}{3}$ is greater than $\frac{1}{4}$, $\left(\frac{1}{3}+\frac{1}{4}\right)$ is greater than $\left(\frac{1}{4}+\frac{1}{4}\right)$; that is, greater than $\frac{2}{4}$, or $\frac{1}{2}$. Again, since $\frac{1}{5}, \frac{1}{6}$, and $\frac{1}{7}$ are all greater than $\frac{1}{8}$, the second group in brackets is greater than $\left(\frac{1}{8}+\frac{1}{8}+\frac{1}{8}+\frac{1}{8}\right)$, or $\frac{4}{8}$, or $\frac{1}{2}$. In the same way, the third group is greater than $\frac{8}{16}$, or $\frac{1}{2}$. And so on. Hence the sum of the series is greater than

$$1+\frac{1}{2}+\frac{1}{2}+\frac{1}{2}+\frac{1}{2}+\frac{1}{2}+\frac{1}{2}+\frac{1}{2}+...,$$

and so the series obviously diverges to infinity, although it does so very slowly.

The condition that the successive terms decrease in size is therefore not sufficient for the convergence of a series in which all the terms are positive. On the other hand, this condition *is* sufficient for the convergence of an 'alternating series' – one in which the terms are alternately positive and negative. This proposition we state without proof. For example, the series

$$1-\frac{1}{2}+\frac{1}{3}-\frac{1}{4}+\frac{1}{5}-\frac{1}{6}+\frac{1}{7}-\frac{1}{8}+\frac{1}{9}-\ldots$$

converges to a finite limit. The value of this limit, to six decimal places,[3] is 0·693147 (actually $\log_e 2$).

*

Certain isolated cases of infinite series had been studied by mathematicians from time to time, but it was not until the nineteenth century that infinite series as a whole, together with the general question of the infinite, began to be treated in a sound, logical manner. In 1851 appeared a small volume entitled *The Paradoxes of the Infinite*. It was the work of Bernard Bolzano, who did not live to see its publication.[4] We can perhaps appreciate what the best minds of the time were struggling with if we look at a few examples taken from Bolzano's book.

Consider the series

$$S=a-a+a-a+a-a+a-a+a-a+\ldots$$

If we group the terms in one way, we have

$$S=(a-a)+(a-a)+(a-a)+(a-a)+\ldots$$
$$=0+0+0+0+\ldots$$
$$=0.$$

On the other hand, if we group the terms in a second way, we can write

$$S=a-(a-a)-(a-a)-(a-a)-(a-a)-\ldots$$
$$=a-0-0-0-0-\ldots$$
$$=a$$

Again, by still another grouping,

$$S=a-(a-a+a-a+a-a+\ldots)$$
$$=a-S.$$

Therefore $2S=a$, or $S=\dfrac{a}{2}$.

Here, then, is an infinite series whose limit is apparently any one of three quantities: 0, or a, or $a/2$. Today, using the definitions of convergence and divergence which we developed in connection with the Zeno paradox, we should say that this series neither converges to a finite limit, nor diverges to infinity. Noting that its sum oscillates between the values 0 and a, we should simply class it as an 'oscillating series' and agree that it has no fixed sum. But in the days before Bolzano ideas of convergence and divergence were not so clearly defined, and such a thing as an oscillating series presented real difficulties. Even Leibnitz – one of the master minds of the seventeenth century, and co-discoverer with Newton of the calculus – was befuddled by this particular series. He argued that since the limits 0 and a are *equally probable*, the correct limit of the series is the average value $a/2$. The method of grouping by which we arrived above at the limit $a/2$ is the work of a mathematician of the early nineteenth century.[5]

Even more startling are the results to be obtained from this series in the special case in which a has the value 1. For example, by actual division we have, for any value of x,

$$\frac{1}{1+x}=1-x+x^2-x^3+x^4-x^5+...,$$

$$\frac{1}{1+x+x^2}=1-x+x^3-x^4+x^6-x^7+...,$$

$$\frac{1}{1+x+x^2+x^3}=1-x+x^4-x^5+x^8-x^9+...,$$

$$\frac{1}{1+x+x^2+x^3+x^4}=1-x+x^5-x^6+x^{10}-x^{11}+...,$$

and so on. Now let x have the value 1. All of the right-hand sides reduce to the *same* number – that is to say, to the 'sum' of the series

$$1-1+1-1+1-1+1-1+...,$$

while the left-hand sides become, respectively, $\frac{1}{2}, \frac{1}{3}, \frac{1}{4}, \frac{1}{5}, ...$ Consequently $\frac{1}{2}=\frac{1}{3}=\frac{1}{4}=\frac{1}{5}=...=1/n$, where n is any natural number! As before, the correct argument is that the series $1-1+1-1+1-1+...$ does not have a fixed sum, but that its sum oscillates between 0 and 1.

Consider still another example of Bolzano's. Let

$$S = 1 - 2 + 4 - 8 + 16 - 32 + 64 - 128 + \dots$$

Then

$$S = 1 - 2(1 - 2 + 4 - 8 + 16 - 32 + 64 - \dots)$$
$$= 1 - 2S.$$

That is,

$$3S = 1, \text{ or } S = \frac{1}{3}.$$

On the other hand, the original series can be written

$$S = 1 + (-2 + 4) + (-8 + 16) + (-32 + 64) + \dots$$
$$= 1 + 2 + 8 + 32 + 64 + \dots,$$

or S diverges to infinity. But again, we can write

$$S = (1 - 2) + (4 - 8) + (16 - 32) + (64 - 128) + \dots,$$
$$= -1 - 4 - 16 - 64 - \dots,$$

or S diverges to negative infinity.

These contradictions are to be explained by the fact that this series is not only an oscillating series, but is one which *oscillates infinitely*. The sum of the first two terms is -1; of the first three, 3; of the first four, -5; and so on through the values 11, -21, 43, -85, ... It is evident that as we go farther and farther out in the series, these partial sums jump from increasingly large positive numbers to increasingly large negative numbers. In a word, the series has no sum.

*

It is perhaps not so surprising that a series which fails to converge to a definite limit can be made to *appear* to converge to a number of different limits. But now consider the series

$$1 - \frac{1}{2} + \frac{1}{3} - \frac{1}{4} + \frac{1}{5} - \frac{1}{6} + \frac{1}{7} - \frac{1}{8} + \dots,$$

which, as has already been pointed out, converges to the finite limit $\log_e 2$, or $0 \cdot 693147$. For simplicity we shall denote this limit by L. Then

$$L = 1 - \frac{1}{2} + \frac{1}{3} - \frac{1}{4} + \frac{1}{5} - \frac{1}{6} + \frac{1}{7} - \frac{1}{8} + \frac{1}{9} -$$
$$\frac{1}{10} + \frac{1}{11} - \frac{1}{12} + \frac{1}{13} - \frac{1}{14} + \frac{1}{15} - \frac{1}{16} + \dots$$

Multiply both sides by 2:

$$2L = 2 - \frac{2}{2} + \frac{2}{3} - \frac{2}{4} + \frac{2}{5} - \frac{2}{6} + \frac{2}{7} - \frac{2}{8} + \frac{2}{9} -$$

$$\frac{2}{10} + \frac{2}{11} - \frac{2}{12} + \frac{2}{13} - \frac{2}{14} + \frac{2}{15} - \frac{2}{16} + \ldots$$

$$= 2 - 1 + \frac{2}{3} - \frac{1}{2} + \frac{2}{5} - \frac{1}{3} + \frac{2}{7} - \frac{1}{4} + \frac{2}{9} - \frac{1}{5} + \frac{2}{11} - \frac{1}{6} + \frac{2}{13} - \frac{1}{7} + \frac{2}{15} - \frac{1}{8} + \ldots$$

Now group terms with the same denominator. Then

$$2L = (2-1) - \frac{1}{2} + \left(\frac{2}{3} - \frac{1}{3}\right) - \frac{1}{4} + \left(\frac{2}{5} - \frac{1}{5}\right) - \frac{1}{6} + \left(\frac{2}{7} - \frac{1}{7}\right) - \frac{1}{8} + \ldots,$$

or

$$2L = 1 - \frac{1}{2} + \frac{1}{3} - \frac{1}{4} + \frac{1}{5} - \frac{1}{6} + \frac{1}{7} - \frac{1}{8} + \ldots$$

But the series on the right is the original series, and its limit is no longer L, but $2L$. Moreover, if the operation of multiplying by two and collecting terms with the same denominator is repeated indefinitely, the series can evidently be summed not only to L and $2L$, but also to $4L$, $8L$, $16L$, ... Here is a real dilemma – an infinite series which converges to a finite limit, $0 \cdot 693147$, yet which can, by proper rearrangements, be made to converge to $1 \cdot 38629$, or $2 \cdot 77259$, or $5 \cdot 54518$, and so on![6]

The difficulty arises from our attempt to apply to *infinite* series the processes of *finite* arithmetic. In finite arithmetic we go on the assumption that we can insert and remove brackets at will, grouping terms in any way we please. In other words, we assume that $A + B + C = (A + B) + C = A + (B + C)$. The contradictory results we obtained above show that this finite operation cannot be applied to infinite series in general.

The question then arises, is it *ever* possible to rearrange and group the terms of a convergent infinite series with the assurance that the limit will not be changed? The answer is yes – provided the series is 'absolutely convergent'. An infinite series is absolutely convergent if not only the series itself converges, but if the series formed by changing all minus signs to plus signs also converges. Thus every convergent series in which all terms are positive is absolutely convergent, and the criterion applies only to series in which there are negative terms.

Let us return for a moment to our original series,

$$1-\frac{1}{2}+\frac{1}{3}-\frac{1}{4}+\frac{1}{5}-\frac{1}{6}+\frac{1}{7}-\frac{1}{8}+\dots$$

If here we change all the minus signs to plus, the series becomes the harmonic series,

$$1+\frac{1}{2}+\frac{1}{3}+\frac{1}{4}+\frac{1}{5}+\frac{1}{6}+\frac{1}{7}+\frac{1}{8}+\dots,$$

which, as we found on page 122, slowly but surely diverges. Consequently our series is not absolutely convergent, and so it is not to be wondered at that we were able – through proper grouping – to make it converge to limits other than $\log_e 2$.

If a series is convergent, but not absolutely convergent, it is said to be 'simply convergent'. The question of rearranging the terms of a simply convergent series was settled in 1854 by the German mathematician Riemann, when he succeeded in proving the following remarkable theorem.[7] *The terms of a simply convergent series can be so rearranged that the limit of the series is any specified finite number, or positive infinity, or negative infinity!*

We conclude this section with four additional examples of the weird results to be had by rearranging and grouping the terms of a simply convergent series. The first two are, essentially, but different forms of the paradox we have just discussed in detail.

Paradox 1. As before, denote by L the value of $\log_e 2$. Then[8]

$$L=1-\frac{1}{2}+\frac{1}{3}-\frac{1}{4}+\frac{1}{5}-\frac{1}{6}+\frac{1}{7}-\frac{1}{8}+\frac{1}{9}-$$

$$\frac{1}{10}+\frac{1}{11}-\frac{1}{12}+\frac{1}{13}-\frac{1}{14}+\frac{1}{15}-\frac{1}{16}+\dots$$

Grouping terms, first by twos and then by fours,

$$L=\left(1-\frac{1}{2}\right)+\left(\frac{1}{3}-\frac{1}{4}\right)+\left(\frac{1}{5}-\frac{1}{6}\right)+$$

$$\left(\frac{1}{7}-\frac{1}{8}\right)+\left(\frac{1}{9}-\frac{1}{10}\right)+\left(\frac{1}{11}-\frac{1}{12}\right)+\dots, \quad (1)$$

and

$$L=\left(1-\frac{1}{2}+\frac{1}{3}-\frac{1}{4}\right)+\left(\frac{1}{5}-\frac{1}{6}+\frac{1}{7}-\frac{1}{8}\right)+\left(\frac{1}{9}-\frac{1}{10}+\frac{1}{11}-\frac{1}{12}\right)+\dots \quad (2)$$

Dividing both sides of equation (1) by 2 we get

$$\frac{1}{2}L = \left(\frac{1}{2} - \frac{1}{4}\right) + \left(\frac{1}{6} - \frac{1}{8}\right) + \left(\frac{1}{10} - \frac{1}{12}\right) + \left(\frac{1}{14} - \frac{1}{16}\right) + \ldots \quad (3)$$

Adding, bracket by bracket, equations (2) and (3),

$$\frac{3}{2}L = \left(1 + \frac{1}{3} - \frac{1}{2}\right) + \left(\frac{1}{5} + \frac{1}{7} - \frac{1}{4}\right) + \left(\frac{1}{9} + \frac{1}{11} - \frac{1}{6}\right) + \left(\frac{1}{13} + \frac{1}{15} - \frac{1}{8}\right) + \ldots$$

$$= 1 - \frac{1}{2} + \frac{1}{3} - \frac{1}{4} + \frac{1}{5} - \frac{1}{6} + \frac{1}{7} - \frac{1}{8} + \frac{1}{9} - \frac{1}{10} + \frac{1}{11} - \ldots$$

Therefore the sum of the series is both L and $\frac{3}{2}L$.

Paradox 2. As in the previous example, denote $\log_e 2$ by L. Then[9]

$$L = 1 - \frac{1}{2} + \frac{1}{3} - \frac{1}{4} + \frac{1}{5} - \frac{1}{6} + \frac{1}{7} - \frac{1}{8} + \ldots$$

Arranging positive terms in one group and negative terms in another,

$$L = \left(1 + \frac{1}{3} + \frac{1}{5} + \frac{1}{7} + \ldots\right) - \left(\frac{1}{2} + \frac{1}{4} + \frac{1}{6} + \frac{1}{8} + \ldots\right). \quad (1)$$

Now certainly

$$0 = \left(\frac{1}{2} + \frac{1}{4} + \frac{1}{6} + \frac{1}{8} + \ldots\right) - \left(\frac{1}{2} + \frac{1}{4} + \frac{1}{6} + \frac{1}{8} + \ldots\right). \quad (2)$$

Adding equations (1) and (2),

$$L = \left[\left(1 + \frac{1}{3} + \frac{1}{5} + \frac{1}{7} + \ldots\right) + \left(\frac{1}{2} + \frac{1}{4} + \frac{1}{6} + \frac{1}{8} + \ldots\right)\right] -$$
$$2\left(\frac{1}{2} + \frac{1}{4} + \frac{1}{6} + \frac{1}{6} + \ldots\right)$$

$$= 1 + \left(\frac{1}{2} + \frac{1}{3} + \frac{1}{4} + \ldots\right) - \left(1 + \frac{1}{2} + \frac{1}{3} + \frac{1}{4} + \ldots\right)$$

$$= 0.$$

In other words, the sum of the series is both L and zero.

Paradox 3. It can be shown that the series,

$$\frac{1}{1.3} + \frac{1}{3.5} + \frac{1}{5.7} + \frac{1}{7.9} + \ldots,$$

is convergent. Call its sum M. We shall 'prove'[10] that M is both 1 and $\frac{1}{2}$

In the first place, the series can be written in the form

$$M = \left(\frac{1}{1} - \frac{2}{3}\right) + \left(\frac{2}{3} - \frac{3}{5}\right) + \left(\frac{3}{5} - \frac{4}{7}\right) + \left(\frac{4}{7} - \frac{5}{9}\right) + \dots$$

To verify this statement, note that the first term reduces to $\frac{3-2}{1.3}$, or $\frac{1}{1.3}$; the second term to $\frac{10-9}{3.5}$, or $\frac{1}{3.5}$; and so on. But if the brackets are now removed, all terms after the first drop out. Therefore $M=1$.

On the other hand, the series can also be written in the form

$$M = \frac{1}{2}\left(\frac{1}{1} - \frac{1}{3}\right) + \frac{1}{2}\left(\frac{1}{3} - \frac{1}{5}\right) + \frac{1}{2}\left(\frac{1}{5} - \frac{1}{7}\right) + \frac{1}{2}\left(\frac{1}{7} - \frac{1}{9}\right) + \dots$$

To verify this statement, note that the first term reduces to $\frac{1}{2} \cdot \frac{3-1}{1.3}$, or $\frac{1}{1.3}$; the second term to $\frac{1}{2} \cdot \frac{5-3}{3.5}$, or $\frac{1}{3.5}$; and so on. If now the brackets are removed, we have

$$M = \frac{1}{2} - \frac{1}{6} + \frac{1}{6} - \frac{1}{10} + \frac{1}{10} - \frac{1}{14} + \frac{1}{14} - \dots$$

Again all terms after the first drop out, so that $M=\frac{1}{2}$.

Paradox 4. To prove that every infinite series, convergent or not, can be summed to any desired number N.
Consider the series

$$a_1 + a_2 + a_3 + a_4 + a_5 + a_6 + \dots$$

We can write

$$a_1 = N + (a_1 - N),$$
$$a_2 = -(a_1 - N) + (a_1 + a_2 - N),$$
$$a_3 = -(a_1 + a_2 - N) + (a_1 + a_2 + a_3 - N),$$
$$a_4 = -(a_1 + a_2 + a_3 - N) + (a_1 + a_2 + a_3 + a_4 - N),$$
$$a_5 = -(a_1 + a_2 + a_3 + a_4 - N) + (a_1 + a_2 + a_3 + a_4 + a_5 - N),$$

and so on indefinitely. Adding these equations, we have

$$a_1+a_2+a_3+a_4+a_5+\ldots$$
$$=N+(a_1-N)-(a_1-N)+(a_1+a_2-N)$$
$$-(a_1+a_2-N)+(a_1+a_2+a_3-N)$$
$$-(a_1+a_2+a_3-N)+(a_1+a_2+a_3+a_4-N)$$
$$-(a_1+a_2+a_3+a_4-N)+\ldots$$

But, now, on the right-hand side of this equation, all terms after the first drop out when the brackets are removed. Consequently the sum of the series on the left-hand side is N.

THE INFINITE IN GEOMETRY

The following paradox appeared some three hundred years ago in Galileo's *Dialogues Concerning Two New Sciences*.[11] It is typical of the confusion which at that time arose from attempts to work with the infinite in geometry.

Take any square $ABCD$ and draw the diagonal BD, as in Figure 76. With B as centre and with radius BC describe the quarter-

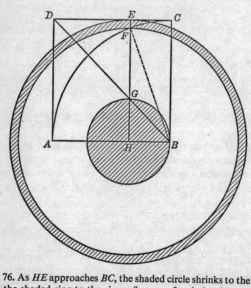

FIG. 76. As HE approaches BC, the shaded circle shrinks to the point B and the shaded ring to the circumference of a circle with radius BC

circle *CFA*. Draw any line *HE* parallel to *BC*, intersecting the quarter-circle at *F* and the diagonal at *G*. With *H* as centre construct circles with radii *HG*, *HF*, and *HE* respectively.

It is not difficult to show that the area of the shaded circle is equal to that of the shaded ring. To do so, note first that triangle *FBH* is a right triangle. Consequently, by the well-known Pythagorean theorem, $\overline{BF}^2 = \overline{HB}^2 + \overline{HF}^2$, or

$$\overline{HB}^2 = \overline{BF}^2 - \overline{HF}^2. \tag{1}$$

But *HE*=*BC*, and, since *BC* and *BF* are radii of the same (quarter) circle, *BC*=*BF*. Hence *HE* and *BF* are equal. Again, *HB*=*HG* since they are also radii of the same circle. We can therefore replace, in equation (1), *BF* by *HE* and *HB* by *HG*, obtaining

$$\overline{HG}^2 = \overline{HE}^2 - \overline{HF}^2. \tag{2}$$

Multiplying both sides of equation (2) by π,

$$\pi.\overline{HG}^2 = \pi.\overline{HE}^2 - \pi.\overline{HF}^2.$$

The left-hand side of this equation represents the area of the shaded circle. The right-hand side, being the difference of the areas of the circles with radii *HE* and *HF*, represents the area of the shaded ring.

Now think of letting *HE* move to the right and approach the position *BC*. As *HE* coincides with *BC* the shaded circle shrinks to the point *B* and the shaded ring shrinks to the circumference of a circle with *HE* (now *BC*) as radius. But since the areas of the shaded circle and the shaded ring are equal for *any* position of *HE*, we must conclude that *a single point is equal to the circumference of a circle!*

Perhaps the solution of this paradox is obvious. It would be more obvious had we not cheated in the statement of the final conclusion. We should have said, 'A single point is equal *in area* to the circumference of a circle.' The circumference of a circle is a one-dimensional curve and can occupy no more area than a point, in spite of the fact that it consists of an infinitude of points. The *areas* of the circumference and of the point, both of them zero, are indeed equal.

*

The infinite in geometry continued to confuse mathematical minds long after Galileo. As late as 1834 an interesting but fallacious proof of the much-discussed 'parallel postulate' of plane geometry was offered, apparently in good faith.[12]. The postulate – assumption, that is – referred to is generally stated as follows: *Through a given point outside a given line, one and only one line can be drawn parallel to the given line*. For centuries it was felt that this postulate could be proved in terms of the other postulates, but all attempts at such proofs were unsuccessful. Mathematicians finally began to suspect that this assumption was as fundamental as the others – as fundamental, for example, as the assumption that between two points one and only one straight line can be drawn. A few of the bolder spirits of the early nineteenth century began to experiment with geometries in which this assumption was replaced by quite different assumptions, and out of the efforts of these pioneers arose the geometries now called 'non-Euclidean'.

The parallel postulate can be stated in a number of forms, each of them equivalent to the others. In the fallacious proof we are about to describe we shall use the form originally given by Euclid: *If two straight lines are cut by a third in such a way that the sum of the interior angles on one side of the third line is less than two right angles, then the two lines, if produced indefinitely, will meet on that side*. That is to say, if the sum of the angles *ABC* and *BCD* of Figure 77(a) is less than two right angles, then *BA* and *CD*, if produced, will ultimately meet.

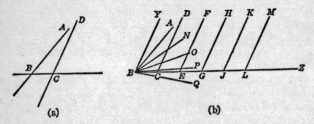

FIG. 77

The hopeful geometer of 1834 set up the figure of diagram (b) to aid him in his proof. Through *B* he drew *BY* parallel to *CD*, and constructed angles *ABN*, *NBO*, *OBP*, and *PBQ* each equal to angle *YBA*. He correctly argued that whatever the size of angle *YBA*, he could, by constructing enough angles *ABN*, *NBO*,.., finally arrive at one whose side (*BQ* in the figure) falls below the line *BZ*. Suppose there are $n-1$ such angles (there are four in the figure). He then marked off $n-1$ segments, CE, EG, GJ, ..., each equal to BC, and through the points of division drew lines EF, GH, JK, ..., each parallel to CD. So far so good. But at this point he began to compare *infinite* areas. He maintained, for example, that the infinite area bounded on two sides by the infinite lines BY and BA is equal to the infinite area bounded on two sides by BA and BN, and that the infinite area bounded on three sides by the line segment BC and the infinite lines BY and CD is equal to the infinite area bounded on three sides by the line segment CE and the infinite lines CD and EF. Or, as he put it, the area YBA is equal to the area ABN, and the area $YBCD$ is equal to the area $DCEF$.

Let us assume for the moment that this argument and similar arguments about infinite areas are valid. The rest of the proof then runs: Area $YBLM$ is equal to n times area $YBCD$, and area YBQ is equal to n times area YBA. But area $YBLM$ is only a part of area YBZ, while area YBZ is in turn only a part of area YBQ. Hence n times area $YBCD$ is less than area YBZ, which in turn is less than n times area YBA. That is to say, $n.(YBCD)$ is less than $n.(YBA)$, or $YBCD$ is less than YBA. But if such is the case, AB must meet CD. For if AB did not meet CD, then $YBCD$ would be equal to the sum of YBA and $ABCD$, and so would be greater then YBA.

The catch in this innocent-looking proof lies, of course, in the fact that the areas involved are infinite. Of two *finite* areas, we can say that the first is less than, equal to, or greater than the second. But two *infinite* areas cannot be compared – we can say only that they are infinite.

*

The remainder of this section will be devoted for the most part to 'limiting curves' – curves which are defined as the limit of an infinite sequence of polygons, that is, of an infinite sequence of figures made up of straight lines. The notion of a limiting curve is not new to any of use who have studied plane geometry. Let us recall briefly how as familiar a curve as the circle can be regarded as the limit of an infinite sequence of regular polygons. (A 'regular polygon' is one with equal sides and equal angles.)

In diagram (a) of Figure 78, a square has been constructed on the line segment AB as diagonal. Diagrams (b), (c), and (d) show, respectively, regular polygons of 2.4 or 8 sides, 2.8 or 16 sides, and 2.16 or 32 sides. Let us denote these successive polygons by P_1, P_2, P_3, and P_4. By continuing indefinitely to double the number of sides we obtain a sequence of polygons, $P_1, P_2, P_3, P_4, P_5, P_6, \ldots$

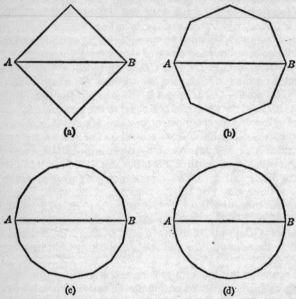

FIG. 78. The circle as the limit of a sequence of regular polygons

Now it is intuitively evident – and it can be proved rigorously – that this sequence of regular polygons approaches, as a limit, the circle whose diameter is *AB*.

*

Care must be used in appealing, as we have just done, to intuition. We must stop to consider three problems in which intuitive arguments lead us wildly astray. The correct solutions of these problems are discussed in the Appendix.

Paradox 1. Consider the isosceles right triangle in Figure 79(a). If each of the equal legs is 1 inch, then, by the Pythagorean theorem, the hypotenuse is $\sqrt{1^2+1^2} = \sqrt{1+1} = \sqrt{2}$ inches. In diagram (b)

the broken line is drawn, beginning at the lower left-hand corner, by going up $\frac{1}{4}$ inch, over $\frac{1}{2}$, up $\frac{1}{2}$, over $\frac{1}{2}$, and up $\frac{1}{4}$. Call this line L_1. In diagram (c) the number of 'steps' is doubled, resulting in the broken line L_2; and in (d) redoubled, resulting in the broken line L_3. Continuing indefinitely the process of doubling the number of

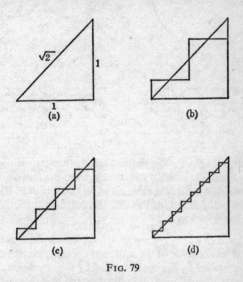

FIG. 79

steps, we obtain a sequence of broken lines, $L_1, L_2, L_3, L_4, L_5, L_6,$ ... This sequence of lines approaches, as a limit, the hypotenuse of the original triangle. Consequently the length of the limiting line is $\sqrt{2}$ inches. True? No, false. What is its length?

Paradox 2. A circle is constructed on a diameter AB, as in Figure 80(a). Call this curve C_1. Now construct a curve consisting of two circles, each of diameter $AB/2$, as in diagram (b). These two circles can be thought of as a single curve, traced as indicated by the arrows. Call this curve C_2. Curves C_3 and C_4 are shown in diagrams (c) and (d). They consist, respectively, of four circles, each of diameter $AB/4$, and of eight circles, each of diameter $AB/8$. Continue indefinitely the process of doubling the number of circles and halving the diameters. The result is a sequence of curves C_1,

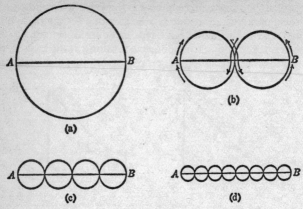

Fig. 80

C_2, C_3, C_4, C_5, C_6, ... The limiting curve, being made up of infinitely small circles, is indistinguishable from the line segment AB. Now recall that in tracing each curve we go from A to B and back to A. Hence the length of the limiting curve is $2.AB$. True? No, false. What is its length?

Paradox 3. In a circle of radius R inscribe a square, as in Figure 81(a), and on each side of the square as diameter construct a semicircle. Denote by C_1 the curve formed by these semi-circles. Now inscribe a regular octagon, as in diagram (b), and denote by C_2 the curve formed by the semicircles constructed on its sides. Continue indefinitely the process of doubling the number of sides of the polygons and constructing semicircles on the sides. The result is a sequence of curves, $C_{,1}$ C_2, C_3, C_4, C_5, C_6, ..., of which C_3 and C_4 are shown in diagrams (c) and (d). The limiting curve, being made up of infinitely small semicircles, is indistinguishable from the original circle of radius R. Its length is therefore $2\pi R$. True? No, false. What is its length?

*

We are now ready to consider some of the so-called 'pathological curves' to be found in mathematics – curves which have been constructed by mathematicians in their attempts to prove or disprove certain intuitive ideas.[13] Each of these curves will be defined, in much the same way that the circle was defined above, as the limit

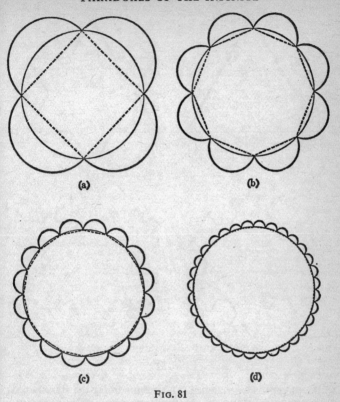

(a)

(b)

(c)

(d)

FIG. 81

of a sequence of polygons, P_1, P_2, P_3, P_4, P_5, P_6, ... In none of the present instances, however, will it be possible, as it was in the case of the circle, actually to draw the limiting curve. We shall have to content ourselves with constructing only the first few polygons of each sequence. The problem of picturing the ultimate curves must be left to our imaginations.

The first item has been dubbed the 'snowflake curve' because of the shape it assumes. P_1, the first polygon of the sequence, is the equilateral triangle of Figure 82(a). Divide each side of this triangle into three equal parts, construct a new equilateral triangle on the middle segment of each side, and do away with lines common

137

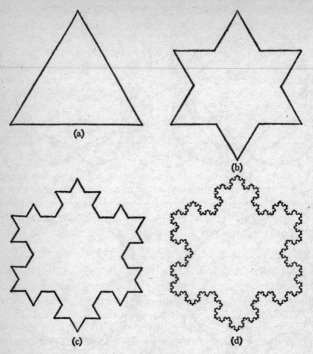

(a)

(b)

(c)

(d)

FIG. 82

to the old and new triangles. This results in P_2, the star-shaped polygon of diagram (b). To get P_3, trisect each side of this polygon, erect a new equilateral triangle on the middle part of each side, and again get rid of lines common to the old polygon and the new triangles. Repeat the same process indefinitely. The result is a sequence of polygons, $P_1, P_2, P_3, P_4, P_5, P_6, \ldots$, of which the third and fifth members are shown in diagrams (c) and (d) respectively of Figure 82.

The limit of this sequence of polygons is a truly remarkable curve: *its length is infinite, yet the area it encloses is finite!* To prove that the area is finite, think first of a circle circumscribed about the original triangle of diagram (a). Then note that at no subsequent stage of the development – as in diagrams (b), (c), and (d) – will the curve ever extend beyond this circle. Now consider the length of the curve. Suppose each side of the original equilateral triangle is 1 unit long. Then the perimeter of P_1 is 3 units. In constructing P_2 we added six lines of length $\frac{1}{3}$ unit and subtracted – by doing away with – three lines of length $\frac{1}{3}$ unit. Net result: we added 1 unit to the perimeter. That is to say, the length of P_2 is $3+1$. In the same way, that of P_3 is $3+1+(\frac{4}{3})$; of P_4, $3+1+(\frac{4}{3})+(\frac{4}{3})^2$; and so on. The perimeter of the limiting curve is therefore the sum of the infinite series

$$3+1+\frac{4}{3}+\left(\frac{4}{3}\right)^2+\left(\frac{4}{3}\right)^3+\left(\frac{4}{3}\right)^4+\ldots$$

It is evident that the successive terms of this series increase in size and that the sum can be made as large as we please by taking a sufficiently large number of terms. Consequently, in accordance with our definition of infinite sum (pp. 118–19), the length of the limiting curve is infinite.

*

A few pages back it was pointed out that a line can occupy no area. This statement is true provided the line is finite in length. But mathematicians have succeeded in constructing a number of limiting curves which *completely fill a given area*! The following curve is one designed by the Polish mathematician, W. Sierpiński.[14]

The first member of the sequence is the polygon P_1, inscribed in a given square as shown in Figure 83(a). The square is then divided into four equal squares, and four polygons, similar to P_1, are joined together to form P_2, as in diagram (b). To get P_3 each of the four squares is divided into four more, and sixteen polygons, again similar to P_1, are joined together as in diagram (c). The same process, repeated, gives the polygon P_4, shown in diagram (d). If the process is continued indefinitely, there results a sequence of polygons, $P_1, P_2, P_3, P_4, P_5, P_6, \ldots$

This sequence of polygons approaches, as a limit, a certain curve. Now it can be rigorously shown that this curve passes through *any* specified point of the square in which it is inscribed. Consequently it must pass through *every* point of the square, *and so* must completely fill it. And if it is not enough of a blow to our intuitions to learn of a one-dimensional curve which fills a two-dimensional square, it might be pointed out that the construction can be generalized to a one-dimensional curve which completely fills

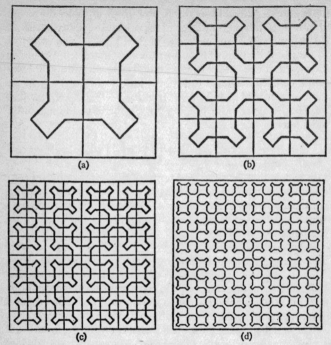

(a)

(b)

(c)

(d)

FIG. 83

a cube in three-dimensional space, or even a 'cube' in a space of any number of dimensions!

*

Before examining the next specimen, we must stop to define clearly what we shall mean by a 'point of intersection' of a curve – that is, a point at which the curve crosses itself.

A point is said to be an 'end point' of a curve if a small circle with the given point as centre always cuts the curve *once*, however small the circle may be. If the arbitrarily small circle about the point cuts the curve *twice*, the point is called a 'general point' of the curve. Finally, if the arbitrarily small circle about the point cuts the curve *more than twice*, the point is said to be a 'point of intersection' of the curve. Thus, in the curve of Figure 84, *P* is an end point, Q is a general point, and R and S are points of intersection. We can agree, can we not, that this definition coincides with our intuitive idea of a point of intersection?

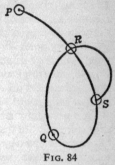

Relying further on intuition, we should undoubtedly say that it is impossible to construct a curve consisting only of points of intersection. That this is not the case was shown by Sierpiński in 1915. His example is constructed as follows:

Divide an equilateral triangle into four congruent equilateral triangles, shade the middle one, and draw in a heavy line as indicated in Figure 85(a). This heavy line is L_1, the first of a sequence of broken lines. Now divide each of the unshaded

FIG. 84

triangles into four congruent triangles, shade the middle one in each case, and draw in the heavy line of diagram (b). This line is L_2,

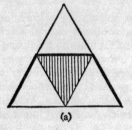

(a)

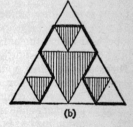

(b)

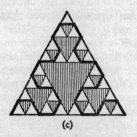

(c)

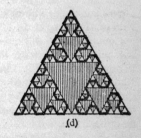

(d)

FIG. 85

141

the second of the sequence. Continue the process indefinitely, at each step dividing the unshaded triangles into four new ones, shading the middle one, and drawing the appropriate heavy line. Diagrams (c) and (d) show the third and fourth members of the resulting sequence of broken lines, $L_1, L_2, L_3, L_4, L_5, L_6, \ldots$

It can be proved that the limit of this sequence is a curve of which every point – with the exception of the vertices of the original triangle – is a point of intersection according to our definition. Finally, if the original triangle is bent out of its plane so as to bring the three vertices together in a single point, then *the curve crosses itself at every point*!

We must admit that this curve is not as easy for the imagination to picture as were the last two curves. The final conclusion will have to be accepted by the non-mathematician on faith. The mathematician, if he so desires, can go to the original source for the proof.[15] Incidentally, both this curve and the area-filling curve are, like the snowflake curve, infinite in length.

*

Let us return, for a moment or two, to the map shown in Figure 57, page 78. In this map there are four countries, each of which touches the other three. For the most part, points on the boundaries between the countries are points common to *two* countries. There are only three points in the figure which are common to *three* countries. In order to be precise, we had better define what we mean by 'a point common to two or more countries'. We can do so in a manner similar to that in which we defined 'point of intersection' in the last example. Thus we shall say that a point is common to two (or more) countries if an arbitrarily small circle about the point as centre includes points of both (or all) countries. Here again is a definition which, if we stop to think of it, coincides with our intuitive ideas.

Common sense will tell us that points common to three countries must be, as they are in the map referred to, isolated points – in other words, that it is impossible for three countries to have a whole *line* of points in common. That this conclusion is false was shown by the Dutch mathematician Brouwer in 1909.[16] Only a mathematician would conceive the weird map we are about to describe, but here it is.

In Figure 86(a) are countries *A*, *B*, and *C*, together with a nice section, D, of unclaimed territory. We shall assume that D is three miles long and one and a half miles wide. Country A first claims all the land in D which lies more than $\frac{1}{2}$ mile from the boundaries of D (diagram b). It is only reasonable, of course, to connect the new territory to the mother country by means of a narrow corridor, but the presence of this corridor will have no effect on the argument to follow. Country B then steps in and takes all of the remaining territory which lies more than $\frac{1}{8}$ mile from the new boundaries of D (diagram c). Country C, not to be outdone, annexes all of the remaining territory which lies more than $\frac{1}{18}$ mile from these still newer boundaries of D (diagram d), But even now there remains quite a bit of unclaimed land, so they begin all over again. A claims all the land which lies more than $\frac{1}{64}$ mile from what are now the boundaries of D; B, all the land more than $\frac{1}{168}$ mile from these new boundaries; C, all the land more than $\frac{1}{486}$ mile from these still newer boundaries; and so on indefinitely.

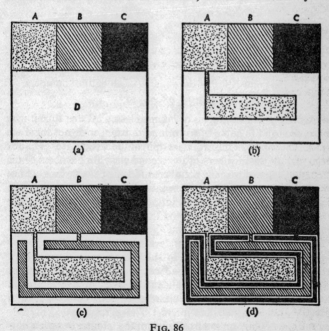

FIG. 86

In the limit, all the original territory D will have been claimed. Furthermore, this can be brought about in a finite length of time by assuming that the first annexation took place in half a year, the second in a quarter of a year, the third in an eighth of a year, and so on. For then the total time in years required to fill the territory completely would be the sum of the infinite series,

$$\frac{1}{2}+\frac{1}{4}+\frac{1}{8}+\frac{1}{16}+\frac{1}{32}+\frac{1}{64}+...,$$

and this, as we have seen earlier in this chapter, comes to one year.

What of the new map of the once unclaimed territory D? It is impossible to draw it, but this much can be said of it. According to our definition of a point common to two or more countries, *every boundary point is a point common to not only two, but to all three of the countries A, B, and C!*

THE ARITHMETIC OF THE INFINITE

At the beginning of this chapter, the class consisting of all natural numbers,

$$1, 2, 3, 4, 5, 6, 7, ...,$$

and the class consisting of the squares of all natural numbers,

$$1, 4, 9, 16, 25, 36, 49, ...,$$

were presented as examples of infinite classes. At that time it may have occurred to some of us to inquire whether or not there are 'more' members of the first class than of the second. It is certainly true that all the members of the second class are members of the first, while there are many members of the first class which are not members of the second. Can it not therefore be said, in spite of the fact that both classes have an infinite number of members, that the infinitude of members in the first case is somehow or other 'greater' than that in the second?

The very problem under consideration was discussed in 1638 by Galileo in his *Dialogues*, a work to which we have already referred.[17] He came to the conclusion that all we can say about the two classes is that each of them is infinite – the relations 'equal', 'greater', and 'less' can be applied to finite classes, but not to infinite classes. There the matter rested until interest in it was reawakened, in 1851, by Bolzano's book on the paradoxes of the infinite. But even Bolzano did not carry his investigations far enough. The possibility of comparing degrees of infinitude was finally realized by Cantor, a German mathematician, in 1873. Out of his work has grown that branch of mathematics called the 'theory of aggregates' – a theory which leads to most extraordinary results.

*

In order to understand Cantor's processes of reasoning, we must begin with *counting* – an operation with which we should be fairly familiar. What do we do when we count the members of a finite class of, say, forty-three objects? It is not enough to say that we point at each member successively and recite, 'One, two, three, ..., forty-two, forty-three'. The ability to perform this operation indicates a highly developed vocabulary of number words. We must go behind the operation of counting by means of words if we are to get at the fundamental process actually involved.

Suppose that you are the leader of an expedition of forty-three people, travelling in an uncivilized country where the vocabulary of number words is limited to 'one', 'two', 'three', 'four', and 'many'. (The existence of savage tribes whose number vocabulary is so limited is well known.) Suppose further that you have gone on ahead to a village where you expect to spend the night, and that you are trying to make the village chief understand that you want food prepared for forty-three people. Assuming that he understands you want food, how will you put over the idea of 'forty-three'? Very likely you will do it by making marks on a piece of paper, or on the ground – one mark corresponding to each member of the party. If a plate of food corresponding to each mark is then prepared, you can be sure that each member of the party will be fed.

Thus the chief, with no word whatever for the number 'forty-three', is able to *count* the number of people in the party, and the number of plates of food. To describe the process in somewhat more precise terms, you set up what is called a 'one-to-one correspondence' between the members of your party and the marks on the paper. The correspondence is 'one-to-one' because corresponding to each person there is one mark, and, conversely, corresponding to each mark there is one person. The chief then sets up a one-to-one correspondence between the marks and the plates of food.

Here is counting in its simplest and most fundamental form – the setting up of a one-to-one correspondence between the members of two classes. The child who counts on his fingers, the Chinese laundryman who reckons his accounts on the abacus, the billiards player who keeps score by means of counters – all of them, consciously or unconsciously, are counting by means of one-to-one correspondences.

Consider one more example. Suppose a theatre contains a certain number of seats – the precise number is immaterial – and suppose the box-office manager wants to know roughly how many people are in the audience. If he notes that every seat is filled and that no one is standing, then he knows that the number is *equal* to the number of seats. In other words, there is a one-to-one correspondence between people and seats. If, on the other hand, some of the seats are empty – if there are seats to which no people correspond – then he knows that the number of people is *less* than the number of seats. Finally, if all the seats are filled and there are some people standing – if there are people to whom no seats correspond – then he knows that the number of people is *greater* than the number of seats.

It should be emphasized that the scheme by which any particular correspondence is set up is of no importance. In order to conclude that two classes of objects have the same number, it is necessary only to exhibit *some sort* of systematic method of establishing a one-to-one correspondence between their members.

*

The natural numbers, 1, 2, 3, 4, 5, ..., are pure abstractions. We go about getting them in essentially the following way. Beginning with some very fundamental and familiar objects – our fingers, let us say – we denote by the symbol '1' the number of any class which can be put into one-to-one correspondence with a single finger. (We should perhaps avoid the word 'number' and use some other word, such as 'plurality' or 'cardinality', but in doing so we should really be begging the question.) In the same way we denote by the symbol '2' the number of any class which can be put into one-to-one correspondence with a pair of fingers, by '5' the number of any class which can be put into one-to-one correspondence with all the fingers of one hand, and so on.

It was Cantor's idea to extend the notion of the sequence of *finite* numbers, 1, 2, 3, 4, 5, 6, ..., to a sequence of *transfinite* numbers. These might be denoted by $A_1, A_2, A_3, A_4, A_5, A_6, ...$ Just as the

finite numbers were associated with certain model finite classes –
we used our fingers – so the transfinite numbers must be associated
with certain model infinite classes. The simplest and most fundamental of all infinite classes seems to be the class consisting of all
the natural numbers. Consequently we denote by A_1 the number
of any class which can be put into one-to-one correspondence with
this particular class. Before attempting to find model classes with
which to associate the other As, let us investigate some classes
which have the number A_1.

Think back to our original problem of the *squares of all natural
numbers*. We can set up a one-to-one correspondence between them
and the natural numbers in the following way:

$$1, \quad 2, \quad 3, \quad 4, \quad 5, \quad 6, \quad 7, \quad ..., \quad n, \quad ...$$
$$\updownarrow \quad \updownarrow \quad \updownarrow \quad \updownarrow \quad \updownarrow \quad \updownarrow \quad \updownarrow \qquad \updownarrow$$
$$1, \quad 4, \quad 9, \quad 16, \quad 25, \quad 36, \quad 49, \quad ..., \quad n^2, \quad ...$$

Now it is true that we cannot show, as we can in the case of two
finite classes, the correspondence which exists between *every* member of the first class and the associated member of the second class,
up to and including the *last* members of each class. There simply
is no last member of either class. On the other hand, it should not
be difficult for our minds to transcend this difficulty. For we know
we are safe in saying that corresponding to every number (n) of the
first class there is a number (n^2) of the second, and, conversely,
corresponding to every number (n^2) of the second class there is a
number (n) of the first. Consequently a one-to-one correspondence
between the two classes does exist, and we are in a position to say
that the class of the squares of all natural numbers has the transfinite number A_1.

Similarly the class of all *even numbers* has the transfinite number
A_1. The correspondence in this case looks like this:

$$1, \quad 2, \quad 3, \quad 4, \quad 5, \quad 6, \quad 7, \quad ..., \quad n, \quad ...$$
$$\updownarrow \quad \updownarrow \quad \updownarrow \quad \updownarrow \quad \updownarrow \quad \updownarrow \quad \updownarrow \qquad \updownarrow$$
$$2, \quad 4, \quad 6, \quad 8, \quad 10, \quad 12, \quad 14, \quad ..., \quad 2n, \quad ...$$

Again, the class of all *odd* numbers has the transfinite number A_1,
for we can write

$$1, \quad 2, \quad 3, \quad 4, \quad 5, \quad 6, \quad 7, \quad ..., \quad n, \quad ...$$
$$\updownarrow \quad \updownarrow \quad \updownarrow \quad \updownarrow \quad \updownarrow \quad \updownarrow \quad \updownarrow \qquad \updownarrow$$
$$1, \quad 3, \quad 5, \quad 7, \quad 9, \quad 11, \quad 13, \quad ..., \quad 2n-1, \quad ...$$

It may already have dawned on some of us that something incredible is going on here. In each of the three examples discussed the class of natural numbers has been put into one-to-one correspondence with a part of itself. In other words, we have been demonstrating that *the whole is equal to part of itself*! This verdict is a direct contradiction of the familiar assumption, first met with in geometry, that *the whole is equal to the sum of its parts and is therefore greater than any of them*. No doubt we have forgotten – if indeed it was ever pointed out to us – that this assumption refers to *finite* magnitudes. We are now working with *infinite* magnitudes for which the assumption, as we can see, is no longer a consistent one.

The whole is equal to part of itself. If ever a conclusion violated common sense, this one is it. But go to the trouble of re-reading the argument which leads to this conclusion. You must admit that there is nothing in the argument itself that violates common sense. As a matter of fact, the principle upon which the entire argument hinges is no more complicated or mysterious than the principle involved in ordinary counting, for the two are identical.

Moreover the conclusion that the whole may be equal to a part of itself can be turned to a useful purpose. At the beginning of this chapter we rather vaguely *described* an infinite class as 'one which cannot be exhausted by counting over any finite period of time.' We can now, with Cantor, *define* an infinite class as 'one which can be put into one-to-one correspondence with a part of itself.'

One more point. The three examples cited lend weight to the argument that the class of natural numbers is the proper class with which to associate A_1, the smallest transfinite number. Note that in each of these examples the natural numbers were thinned out, yet the number of members of the resulting classes remained the same. The thinning-out process can be carried on indefinitely, and always with the same result. Thus all the classes

$$4, 8, 12, 16, 20, 24, \ldots, 4n, \ldots,$$
$$8, 16, 24, 32, 40, 48, \ldots, 8n, \ldots,$$
$$100, 200, 300, 400, 500, \ldots, 100n, \ldots,$$
$$10^{100}, 2.10^{100}, 3.10^{100}, \ldots, n.10^{100}, \ldots,$$

have the same transfinite number as the class of all natural numbers.

*

Now let us turn our attention to the problem of finding an infinite class whose number is greater than A_1. One possibility that may suggest itself is the class of *all rational numbers*.

We recall from algebra that a rational number is defined as one which can be written as the quotient of two whole numbers. For example, 2/3, −5/8, and 4/7 are rational numbers. It is at once evident that the class of rational numbers includes the class of natural numbers, for 1 can be expressed as 1/1, 2 as 2/1, 3 as 3/1, and so on. Again, all ordinary decimals are rational numbers, for such a decimal as 3·579 can be written as 3579/1000. Finally, all repeating decimals are rational numbers, for 0·3333333 ... can be written as 1/3, 0·3454545 ... as 19/55, 2·4272727 ... as 267/110, and so on. For convenience we shall restrict our attention to positive rationals. Consequently we shall be considering all numbers of the form p/q, where p and q are natural numbers.

An important property of the rational numbers lies in the fact that they are 'dense'. By this it is meant that between any two rational numbers there are infinitely many other rational numbers. For example, between 0 and 1 we can point to the numbers

$$\frac{1}{2}, \frac{2}{3}, \frac{3}{4}, \frac{4}{5}, \frac{5}{6}, \dots, \frac{n}{n+1}, \dots;$$

between 0 and $\frac{1}{2}$, the numbers

$$\frac{1}{3}, \frac{2}{5}, \frac{3}{7}, \frac{4}{9}, \frac{5}{11}, \dots, \frac{n}{2n+1}, \dots;$$

between 0 and $\frac{1}{4}$, the numbers

$$\frac{1}{5}, \frac{2}{9}, \frac{3}{13}, \frac{4}{17}, \frac{5}{21}, \dots, \frac{n}{4n+1}, \dots;$$

and so on. Because of this property we might well expect the transfinite number of rational numbers to be greater than A_1. Cantor showed that this is *not* the case. His proof runs as follows:

The class of all rational numbers can be arranged as shown in Figure 87. Note that in each horizontal row the successive denominators are 1, 2, 3, 4, 5, 6, ..., while all the numerators in the first row are 1, all those in the second row are 2, all those in the third row are 3, and so on. Note also that every fraction in which the numerator and denominator have a common factor has been enclosed in brackets. If these particular fractions are deleted, then every rational number appears once and only once in the array. Following the path indicated by the arrows, a one-to-one correspondence can be set up between the natural numbers and the

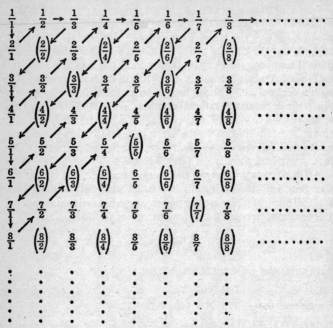

FIG. 87. 'Counting' the rational numbers

rational numbers by pairing with 1 the fraction 1/1, with 2 the fraction 2/1, with 3 the fraction 1/2, with 4 the fraction 1/3, with 5 the fraction 3/1, and so on, as indicated in the following scheme:

$$
\begin{array}{ccccccccccccc}
1, & 2, & 3, & 4, & 5, & 6, & 7, & 8, & 9, & 10, & 11, & 12, & 13, \ldots \\
\updownarrow & \updownarrow & \updownarrow & \updownarrow & \updownarrow & \updownarrow & \updownarrow & \updownarrow & \updownarrow & \updownarrow & \updownarrow & \updownarrow & \updownarrow \\
\dfrac{1}{1}, & \dfrac{2}{1}, & \dfrac{1}{2}, & \dfrac{1}{3}, & \dfrac{3}{1}, & \dfrac{4}{1}, & \dfrac{3}{2}, & \dfrac{2}{3}, & \dfrac{1}{4}, & \dfrac{1}{5}, & \dfrac{5}{1}, & \dfrac{6}{1}, & \dfrac{5}{2} \ldots
\end{array}
$$

There may be some objection to the order, or rather the lack of order, in this set-up. It may be pointed out that the example involving the correspondence between the natural numbers and their squares was more convincing, in that the nth square could be expressed in terms of the nth natural number – that is, as n^2. In the present example there is no such simple relationship between the nth natural number and the nth rational number. Granted. But anyone who raises this objection is forgetting an important point which was emphasized earlier – namely, that the particular way in which the correspondence is set up is immaterial. The important thing is simply to exhibit some sort of systematic way of pairing the members of the two classes. A moment's reflection will make it evident that this has certainly been done here. We first arranged the rational numbers in an array in which every number appeared once and only once. We then indicated the path which should be followed in pairing each rational number with a natural number. If we name any rational number at random we can, by going out far enough in the scheme, find the natural number which corresponds to it. Again, if we select a natural number at random we can, in the same way, find the rational number which corresponds to it. To every rational there corresponds one and only one natural, and to every natural there corresponds one and only one rational. The correspondence is therefore one-to-one, and the fact that the class of positive rational numbers has the transfinite number A_1 has been established.

*

Our first attempt to find an infinite class whose number is greater than A_1 has been a vain one. No doubt a few of us are beginning to suspect that *all* infinite classes have the number A_1. Again Cantor was able to show how wrong our guesses – based on intuition – may be, for he succeeded in proving[18] that the transfinite number of the class of *all real numbers* is greater than A_1.

For our purposes we may define a real number as any number which is not imaginary – that is to say, which does not involve $\sqrt{-1}$. Thus the class of all real numbers includes not only the class of all rational numbers, but the class of all irrational numbers as well. Examples of irrational numbers are $\sqrt{2}$, $\sqrt[3]{5}$, π, e, and $\log_e$ 10. Such a number as $\sqrt{2}$ arises in geometry when we attempt to measure the hypotenuse of a right triangle each of whose other two sides is 1 unit long (see p. 135); the number $\sqrt[3]{5}$ can be interpreted as a solution of the algebraic equation $x^3 = 5$; the number π is indispensable in the measurement of the circle, as is the number e in the study of the calculus; and so on.

Before getting into Cantor's proof we had better make three observations. The first of these concerns the precise meaning of 'greater than' as applied to transfinite numbers. Recall the finite problem of the theatre and the audience. There we found we could say that the number of people is greater than the number of seats if there are any people standing – that is to say, if there are any people to whom no seats correspond. Incidentally, we need point to only *one* person standing in order to draw this conclusion. We shall use this same criterion in connection with infinite classes. Suppose we are attempting to set up a one-to-one correspondence between two infinite classes. If we find that to every member of the first class there corresponds a member of the second class, but that there are some members (there need be only one) of the second class to which no member of the first class corresponds, then we can conclude that the transfinite number of the second class is greater than that of the first.

The second observation concerns the possibility of finding a uniform representation for all real numbers. Such a representation is furnished by non-terminating decimals. It was pointed out, in connection with the definition of a rational number (page 149), that any repeating decimal is equivalent to a rational number. Conversely, every rational number is equivalent to a repeating decimal. Thus 1/3 can be expressed 0·333333333 ..., 10/9 as 1·111111111 ..., 63/65 as 1·145454545 ..., and 10/7 as 1·42857142857 ... Even such numbers as 3 and 5/2, which we would ordinarily write as the terminating decimals 3·0 and 2·5 can be written in non-terminating form as 2·999999999 ... and 2·499999999 ... Real numbers which are not rational – that is to say, irrational numbers – can be expressed as non-terminating decimals which do not repeat. Thus $\sqrt{2}$ can be expressed as 1·414213562 ..., π as 3·141592654 ..., e as 2·718281828 ... (here the group '1828' appears to repeat, but does not do so after the first nine decimal places), logs 10 as 2·302585093 ..., and so on.

Our third observation is to the effect that we shall restrict our attention to the real numbers between 0 and 1. We shall show later how to set up a correspondence between these numbers and all positive real numbers.

And now for Cantor's proof. The gist of the argument is as follows. We shall *assume* that a one-to-one correspondence has been established between the natural numbers and the real numbers from 0 to 1. We shall then exhibit a number, also between 0 and 1, which cannot possibly be included in the scheme – in other words, a real number to which no natural number corresponds.

In the assumed set-up let us denote the successive digits of the first real number, expressed as a non-terminating decimal, by a_1, a_2, a_3, a_4, a_5, ..., those of the second number by b_1, b_2, b_3, b_4, b_5, ..., and so on. Then the correspondence between the numbers will look like Figure 88. Remember we are assuming that *all* real numbers between 0 and 1 appear in the array at the right. We now construct a number, denoted by $\cdot z_1 z_2 z_3 z_4 z_5 z_6 z_7$ ..., in the following way. Proceeding along the diagonal line of the figure, we choose z_1 different from a_1, z_2 different from b_2, z_3 different from c_3, z_4 different from d_4, z_5 different from e_5, and so on.

Now this new number obviously lies between 0 and 1. Furthermore it is not to be found anywhere in the array of real numbers, for it differs from the first number in the first decimal place, from

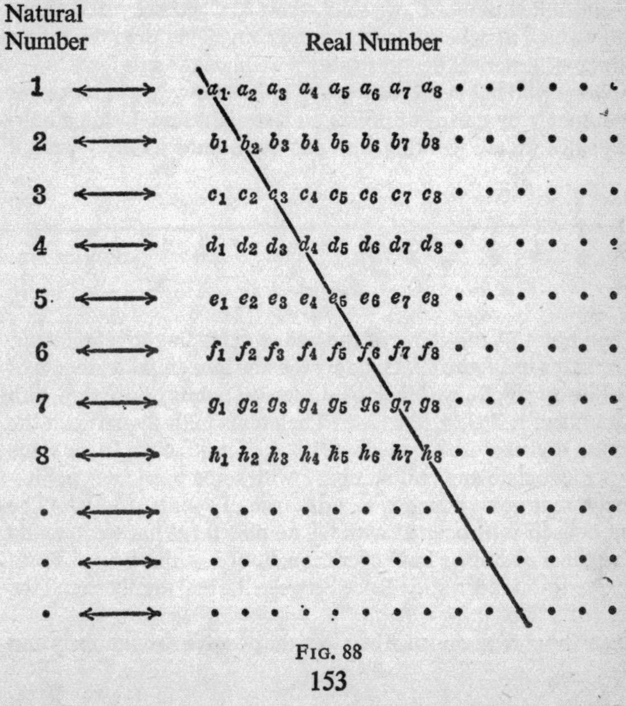

Natural
Number   Real Number

FIG. 88
153

the second number in the second decimal place, from the third number in the third decimal place, and so on. Consequently to this new number corresponds *no* natural number in the left-hand column. It follows that our assumption that the one-to-one correspondence could be established is false, and that *the transfinite number of the class of all real numbers between 0 and 1 is greater than A_1.*

We shall denote this new transfinite number by the symbol C. We might be tempted to identify it with A_2, the transfinite number next greater than A_1. That C and A_2 are the same is perhaps true, yet no one has ever been able to prove it. In other words there *may* be a transfinite number greater than A_1 and at the same time less than C. The question is still an open one. (See note in Appendix, page 221.)

*

In order to show that the class of *all positive real numbers* also has the transfinite number C, we shall resort to a geometrical demonstration which may be somewhat more convincing than the rather abstract arithmetical demonstrations we have used so far.

Anyone who has ever seen a graph knows how we can represent real numbers by means of points on a straight line. Using a half-line – since we are working with positive numbers only – we call

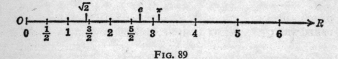

FIG. 89

the end point O and think of the line as extending indefinitely to the right, as in Figure 89. Dividing the line into equal segments of an arbitrary length, we label the successive points of division with the numbers 1, 2, 3, 4, 5, 6, 7, ... The points midway between the points of division are labelled $\frac{1}{2}, \frac{3}{2}, \frac{5}{2}, \frac{7}{2}, \frac{9}{2}, \ldots$, and so on. In the same way we associate any real number r with some particular point – namely, that point which is at a distance of r units from O. (The point O itself is associated with the number 0.) What we actually do, whether or not we have ever thought of it in this way, is to set up a one-to-one correspondence between the real numbers and the points of a line.

Once the correspondence between the positive real numbers and the points of the half-line OR is established, the problem of showing that *all positive real numbers can be put into one-to-one correspondence with the real numbers between 0 and 1 reduces* to the problem of showing that *all points of the half-line OR can be put into one-to-one correspondence with the points of the interval from 0 to 1.*

The second of these problems is handled in the following way. On the half-line OR construct the rectangle $OLMN$ as shown in Figure 90. Make the length OL of the rectangle 1 unit long; its height is immaterial. Let P be any point of OL. At P erect a line perpendicular to OL. This line meets the diagonal OM at S. Draw NS, and extend it to meet OR at Q. The point P of OL is thus

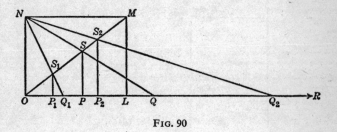

FIG. 90

paired with the point Q of OR. In exactly the same way, P_1 is paired with Q_1, P_2 with Q_2, and so on. Conversely, given the point Q of OR, the corresponding point P of OL can be located by drawing QN and dropping a perpendicular to OL from the point S at which QN meets OM. The correspondence is obviously one-to-one, for to every point of OL corresponds one and only one point of OR, and to every point of OR corresponds one and only one point of OL.

Our argument not only proves that the class of all positive real numbers has the transfinite number C, but uncovers a new and startling paradox as well. For we have just shown that *there are no more points in a line of infinite length than in a line segment one unit long!*

*

If we set out to look for a transfinite number greater than C, it might occur to us to investigate the class consisting of all the points of a plane. For surely there are more points in a *plane* than in a *line.* But are there? We ought by this time to have learned to be suspicious of our intuitive guesses.

In order to work with points of a plane we extend the relationship between *single* real numbers and points of a *line* to a similar relationship between *pairs* of real numbers and points of a *plane.* Thus in Figure 91 the point P_1, which is 2 units from OY and 3 units from OX, can be represented by the pair of real numbers

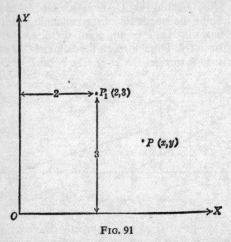

FIG. 91

(2,3). Similarly, any point P of the plane can be made to correspond to the pair of real numbers (x, y), in which the first number represents the distance of P from OY in the direction of OX, and the second number the distance of P from OX in the direction of OY. The correspondence is evidently one-to-one, for to every point there corresponds one and only one pair of numbers, and to every pair of numbers there corresponds one and only one point. (It is to be noted that since we are confining our attention to positive real numbers, we are restricted to those points which lie to the right of OY and above OX. The representation of points elsewhere in the plane involves the use of negative numbers.)

Now in Figure 92 consider the square $OLMN$, each side of which is one unit long, and the line segment OS, which is also one unit long. We shall show that to every point P of the square there corresponds a point Q of the unit segment.

Let the point P be represented by the pair of numbers (x, y). Since both x and y are less than 1, they can be expressed (see page 153) as the non-terminating decimals,

$$x = \cdot x_1 x_2 x_3 x_4 x_5 x_6 x_7 x_8 \ldots,$$
$$y = \cdot y_1 y_2 y_3 y_4 y_5 y_6 y_7 y_8 \ldots$$

Now from the successive digits of the numbers x and y let us form the number

$$z = \cdot x_1 y_1 x_2 y_2 x_3 y_3 x_4 y_4 x_5 y_5 \ldots$$

(For example, if $x = \cdot 3427427427 \ldots$ and $y = \cdot 6129846035 \ldots$, then $z = \cdot 36412279482476402375 \ldots$) This number certainly has a value between 0 and 1, and can therefore be represented by a point Q of

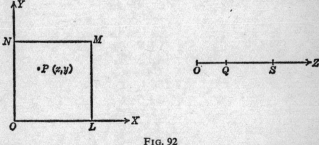

FIG. 92

the line segment OS. That is to say, given the point P of the unit square, we can determine x and y, and therefore z, and so can locate the corresponding point Q of the unit segment.

Our argument shows very simply that there are no more points in the unit square than in the unit line.[19] The proof can be extended to show that there are as many points in the unit line as in a plane of infinite extent. Indeed, if we wish to carry the argument still further, we can show that there are as many points in a line one inch long as in all of three-dimensional space. Finally, if we wish to go to an extreme which appears to be utterly ridiculous, we can prove that *there are as many points in a line a billionth of an inch long as there are in the whole of a space of 4, 5, 6, ..., n, or even A_1 dimensions!*[20]

*

The problem of finding a class whose transfinite number is greater than C is somewhat more complicated than the problems we have so far tackled. And most of these have no doubt been complicated enough. We shall therefore content ourselves with the statement that it has been proved that there is an infinitude of transfinite numbers, and that they can be arranged in order of increasing size. Just as there is no last – or largest – natural number, so there is no last – or largest – transfinite number.[21]

Before leaving the transfinite numbers entirely, however, let us look at the results of certain arithmetical operations on them. If n is any finite natural number, and if A_1 and C are the transfinite numbers with which we are acquainted, then, incredible as it may seem, the following conclusions can be shown to be true.

$$A_1 + n = A_1,$$
$$A_1 + A_1 = A_1,$$
$$n \cdot A_1 = A_1,$$
$$A_1 \cdot A_1 = A_1,$$
$$(A_1)^n = A_1,$$
$$(2)^{A_1} = (A_1)^{A_1} = C,$$
$$C + n = C,$$
$$C + A_1 = C,$$
$$C + C = C,$$
$$n \cdot C = C,$$
$$C \cdot C = C,$$
$$(C)^n = C,$$
$$(C)^{A_1} = C,$$
$$(2)^C = (C)^C = \text{a new transfinite number,}$$

and so on.

*

We need not think, in closing this chapter, that we have seen the last of the infinite and its vagaries. In the following chapters – particularly the next two – there is ample evidence of the fact that the notion of infinity is one of the greatest enemies of the mathematician's peace of mind.

CHAPTER EIGHT

Paradoxes in Probability

IN 1654 the Chevalier de Méré, gambler and amateur mathematician, proposed to Blaise Pascal a problem concerning the division of stakes in a game of dice. Pascal communicated the problem to Fermat, and from the correspondence[1] between these two men arose what has subsequently become the modern theory of probability. Thus did a simple gambler's problem give birth to a powerful technique which constitutes the very foundation of mathematical statistics, and, through statistics, of much of the mathematics of economics and industry.

Most mathematical theories, in the course of their development, have suffered severely from what might well be called 'growing pains'. The theory of probability is no exception. Numerous contradictions have arisen and have led to bitter controversies over concepts of the most fundamental nature. It is these contradictions with which we shall be concerned. In some instances we may not be able to arrive at an entirely satisfactory solution of the difficulty. A few of the problems involve high-powered ideas into which we shall not have time to go in detail, while others are still in dispute among even the better mathematicians.

In order to see how easily misunderstandings may arise, consider the type of problem originally discussed by Pascal and Fermat. Suppose that two players, A and B, contribute equally to a stake of £12. They agree that the first player who makes 3 points shall win the entire stake. After A has won 2 points, and B has won 1, they agree to stop. How should the stake be divided?

Offhand this problem appears to be very simple. We may well argue that since A has twice as many points as B, A's share should be twice B's. That is to say, A should take £8, and B £4. But now suppose they were to play the next point – the one they have agreed not to play. If A were to win this point, the whole stake of £12 would be his. If he were to lose, the score would then be 2 to 2, and they would split the £12 evenly. Thus A is sure of getting £6 anyway. And assuming that he has an even chance of winning the next point, his share of the remaining £6 should be half that amount. In other words, A should take £9, and B £3.

It is not difficult to see that the second solution is correct if A and B are to stick to their original agreement concerning the winning of the stake. Had they agreed to divide the stake in proportion to their scores at any stage of the game, the correct solution would of course be the first one.

But we must not jump too quickly into the midst of difficulties. We had better spend a few moments discussing some of the basic principles of probability, in order to prepare ourselves for the troubles that lie ahead.

*

Laplace, an eminent French mathematician of the late eighteenth and early nineteenth centuries, once described the theory of probability as nothing but 'common sense reduced to calculation'. Let us see to what extent the following anecdote justifies this description.

Two college students are trying to decide how to pass an evening. They finally agree to let their decision rest on the toss of a coin. Heads, they go to the cinema. Tails, they go out for a beer. And if the coin stands on edge, they study!

This story is not as trivial as it may seem, for we can learn much from it. Common sense, basing its judgement on past experience, tells us that the boys will be spared the necessity of studying. In other words we know instinctively that the coin will not stand on edge, but that it will come to rest with either heads or tails showing. Moreover, if the coin is a fair one – if it doesn't have heads, say, on both sides – we are morally certain that the possibility of heads and the possibility of tails are *equally likely* possibilities.

Now the theory of probability is based on the assumptions we make concerning such questions as these: What is the probability that the coin will stand on edge? What is the probability that it will show either heads or tails? What is the probability that it will show heads? What is the probability that it will show tails?

In order to discuss these questions in mathematical terms, it is necessary to assign numerical values to the various probabilities involved. Suppose for the moment that we denote by p the numerical value of the probability that the coin will show heads. Since it is equally likely that the coin will show tails, the probability of tails must also have the value p. But we are *certain* that the coin will show *either* heads *or* tails. Hence $2p$ must have the value of certainty – of the probability that an event which is bound to occur will occur. We can choose for certainty any value we please. It is customary, and convenient, to choose the value 1. That is to say, we assume that $2p=1$. Then the probability that the coin will show heads is $\frac{1}{2}$; that it will show tails, $\frac{1}{2}$; and that it will show either heads or tails, $\frac{1}{2}+\frac{1}{2}$, or 1.

We can generalize our definition of the measure of probability in the following way: *Suppose that the number of ways in which a certain event can happen is h, and that the number of ways in which it can fail to happen is f. Suppose further that the ways in which the event can happen or fail to happen are all equally likely. Then the probability that the event will happen is $h/(h+f)$, the probability that it will fail to happen is $f/(h+f)$, and the probability that it will happen or fail to happen is $h/(h+f)+f/(h+f)=(h+f)/(h+f)=1$.* For example, suppose a single marble is to be drawn from a box containing 3 red marbles and 7 white marbles. Then the probability of drawing a red marble is $\frac{3}{10}$, that of drawing a white marble is $\frac{7}{10}$, and that of drawing a red marble *or* a white marble is $\frac{3}{10}+\frac{7}{10}$, or 1.

In our example of the coin, the only question we have left unanswered is that which concerns the probability that the coin will stand on edge. We have agreed that the coin cannot stand on edge, but that it must fall in either of two ways – with heads showing or with tails showing. That is to say, the number of ways in which the coin can stand on edge is 0, and the number of ways in which this event can fail to happen is 2. Therefore the probability that the coin will stand on edge is $\frac{0}{2}$, or 0. The same reasoning can be applied to the problem of the box of red and white marbles. Since there is no possible way of drawing a marble of any colour other than red or white – say black – then the probability of drawing a black marble is 0.

To summarize our findings briefly, we shall say that the probability of the occurrence of an impossible event is 0, the probability of the occurrence of an event which is certain to occur is 1, and the probability of the occurrence of a doubtful-but-nevertheless-possible event is some fraction between 0 and 1.

Now let us consider a few straightforward examples concerning throws with dice. These examples will serve not only to fix in our minds the ideas just presented, but will also introduce us to one or two elementary short-cuts which we may find useful later on.

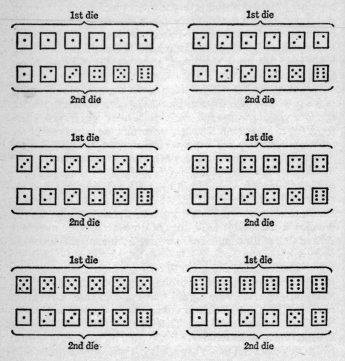

FIG. 93. There are 36 possible throws with a pair of dice

What is the probability of throwing a two with one throw of a single die? Since the die has six faces, any one of which may turn up, there is a total of six equally likely ways in which the desired event can occur or fail to occur. There is only one way in which it can occur. Therefore the probability is $\frac{1}{6}$.

What is the probability of throwing a two *or* a three with one throw of a single die? Again there is a total of six ways in which the proposed event can occur or fail to occur. There are two ways in which it can occur. Therefore the probability is $\frac{2}{6}$, or $\frac{1}{3}$. The same result can be arrived at in another way. Noting that the probability of a two is $\frac{1}{6}$, and that the probability of a three is $\frac{1}{6}$, we can say that the probability of a two *or* a three is $\frac{1}{6}+\frac{1}{6}$, or $\frac{1}{3}$. This argument can be extended to the following general principle: If *p_1, p_2, p_3, ..., p_n are the respective probabilities of n mutually exclusive events, then the probability that one of the events will occur is the sum of these probabilities, or $p_1+p_2+p_3+...+p_n$.* (The throwing of a two and the throwing of a three are mutually exclusive events since they cannot both happen in one throw with a single die.)

What is the probability of throwing two ones with one throw of a pair of dice? Since every number on the first die can be associated with 6 numbers on the second, and since there are 6 numbers on the first die, the dice can fall in any one of $6 . 6 = 36$ possible ways. This point is illustrated in detail in Figure 93. Of the 36 possibilities, only 1 is favourable – that in which the number on both dice is one. Hence the probability of throwing 'snake-eyes' is $\frac{1}{36}$. Note that this same result could have been obtained by the following argument. The probability that the first die turns up a one is $\frac{1}{6}$, and the probability that the second die turns up a one is also $\frac{1}{6}$. Therefore the probability that both dice turn up ones is $(\frac{1}{6})$ $(\frac{1}{6})$, or $\frac{1}{36}$. In general we can say that *if p_1, p_2, p_3, ..., p_n are the respective probabilities of n independent events, then the probability that all n of the events will occur at once is the product of these probabilities, or (p_1) (p_2) (p_3) ... (p_n).* (The throwing of a one with the first die and the throwing of a one with the second die are independent events because the first had no effect whatever on the second.)

One more general point. Note that *if p is the probability that a certain event will occur, then the probability that it will fail to occur is $1-p$.* Thus, in our last problem, if $\frac{1}{36}$ is the probability of throwing 2 ones with a pair of dice, we can conclude that the probability of not throwing 2 ones is $1-(\frac{1}{36})$, or $\frac{35}{36}$. This result is easily verified by noting that if the event can happen in only 1 of 36 possible ways, then it can fail to happen in 35 of those 36 ways.

*

We are now ready to begin our excursion into the paradoxes of probability. Our first example is of some historical interest in that D'Alembert, a first-rate French mathematician of the eighteenth century, failed to solve it correctly.

In two tosses of a single coin, what is the probability that heads will appear at least once?

Noting that heads on the first toss can be associated with either heads or tails on the second toss, and that tails on the first toss can similarly be associated with either heads or tails on the second

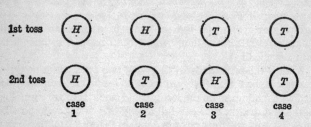

FIG. 94. The four possible results of two tosses of a single coin

toss, the total number of possible cases is 4, as indicated in Figure 94. Of these 4, the first 3 are favourable in that they contain at least 1 head. Therefore the desired probability is $\frac{3}{4}$.

When this problem was proposed to D'Alembert in 1754, he argued as follows:[2] There are only 3 cases: heads on the first throw, or heads on the second throw, or heads not at all. Now 2 of these 3 cases – the first 2 are favourable. Therefore the desired probability is $\frac{2}{3}$.

It does not take long to see why this second solution is wrong. D'Alembert's first case included the first *two* cases shown in Figure 94. In other words, 'heads on the first throw' meant, to D'Alembert, 'heads on the first throw regardless of what happens on the second throw', whereas 'heads on the second throw' meant 'tails on the first throw followed by heads on the second throw.' It is true of D'Alembert's system that one of his three possibilities must occur, and that the possibilities are mutually exclusive. The trouble is that they are not equally likely. It is evident at once from Figure 94 that 'heads on the first throw regardless of what happens on the second throw' (cases 1 and 2) is twice as likely as 'tails on the first throw and heads on the second' (case 3).

The solutions of the following two problems, which involve difficulties similar to those just discussed, will be found in the Appendix.

Paradox 1. Three coins are tossed at once. What is the probability that all three come down alike – that is to say, that all three are either heads or tails ?[3]

(a) We can say with assurance that of the three coins tossed, two of them must come down alike – both heads or both tails. What of the third coin? The probability that it is heads is $\frac{1}{2}$; that it is tails, also $\frac{1}{2}$. In either case the probability that it is the *same* as the other two is $\frac{1}{2}$. Consequently the probability that all three are alike is $\frac{1}{2}$.

(b) But now suppose we use an argument involving the multiplicative and additive principles discussed earlier. For the moment let us fix our attention on heads. The probability that the first coin is heads is $\frac{1}{2}$; that the second is heads, $\frac{1}{2}$; and that the third is heads, $\frac{1}{2}$. Hence the probability that all three are heads is $(\frac{1}{2})$ $(\frac{1}{2})$ $(\frac{1}{2})$, or $\frac{1}{8}$. In exactly the same way, the probability that all three are tails is $\frac{1}{8}$. Therefore the probability that all three are alike – either heads or tails – is $\frac{1}{8} + \frac{1}{8}$, or $\frac{1}{4}$.

Which result are we to accept, $\frac{1}{2}$ or $\frac{1}{4}$?

Paradox 2. Peter and Paul (favourite characters with writers on probability) *play a game of marbles. Peter has two marbles, Paul one. They roll to see which comes nearer some fixed point – say a stake set in the ground. Assuming that they are equally skilful, what is the probability of Peter's winning?*

(a) Since the players are equally skilful, all 3 marbles have the same chance of winning. But 2 of the 3 marbles are Peter's. Therefore the probability that Peter will win is $\frac{2}{3}$.

(b) There are 4 possible cases. Of Peter's 2 marbles, both can be better than Paul's, or the first can be better and the second worse, or the second can be better and the first worse, or both can be worse. Of these 4 cases, the only one which makes Peter lose is the last – that in which both of his marbles are worse than Paul's. Hence the probability that Peter will win is $\frac{3}{4}$.

Which results are we to accept, $\frac{2}{3}$ or $\frac{3}{4}$?

*

A number of paradoxes were discussed by the French mathematician J. Bertrand in his *Calcul des probabilités*, a scholarly treatise which appeared in 1889. One of these in particular has been used as an illustrative example in almost every subsequent textbook on probability. Generally known as 'Bertrand's box paradox', it runs as follows.[5]

Three boxes are identical in external appearance. The first box contains two gold coins, the second contains two silver coins, and the third contains a coin of each kind – one gold and one silver. A box is chosen at random. What is the probability that it contains the unlike coins?

This problem appears to be straightforward enough. There are 3 possible cases: gold-gold, silver-silver, and gold-silver. Since the boxes are identical in appearance, the 3 cases are equally likely. And of the 3 only 1 – the last – is favourable. Therefore the desired probability is $\frac{1}{3}$.

Granting that the solution just suggested is correct (which it is), what is to be done with the following argument? Suppose we choose a box and remove one of the two coins in it. Regardless of whether this coin is gold or silver – it is not at all necessary to examine it – there are only 2 possible cases: the remaining coin is either gold or silver. In other words, it is either like or unlike the coin that has been removed. Of the 2 specified possibilities, 1 is favourable. Hence the probability that the second coin is unlike the first is $\frac{1}{2}$. We are therefore led to the startling conclusion that the removal of one coin from one of the boxes raises the desired probability from $\frac{1}{3}$ to $\frac{1}{2}$! There is certainly something wrong with this argument, for the mere removal of one coin does not increase our knowledge of the nature of the remaining coin.

A number of solutions of this paradox have been proposed, and special techniques have been devised to take care of this difficulty and of similar difficulties.[6] Let us see if Bertrand's own ideas about the matter are sufficiently satisfactory.

Bertrand maintained that, once one coin has been removed, the possibilities subsequently specified are not equally likely. That is to say, if we suppose that the first coin removed is gold, the second coin is less likely to be silver than gold. Why? Well, for simplicity denote the box containing the two gold coins by B_{gg}, that containing the two silver coins by B_{ss}, and that containing one of each by B_{gs}. Then if the first coin removed is gold, it must have come from either B_{gg} or B_{gs} – it obviously could not have come from B_{ss}. Now the chance that the first coin removed from B_{gg} is gold is evidently 1, or certainty; whereas the chance that the first coin removed from B_{gs} is gold is $\frac{1}{2}$. Hence, if a gold coin has been drawn, it is less likely that it came from B_{gs} than from B_{gg}. Consequently the second coin is less likely to be silver than gold. In the same way, if the first coin is silver, it is less likely that it came from B_{gs} than from B_{ss}, so that in this case the second coin is less likely to be gold than silver. Regardless, then, of what the first coin is, the second coin is less likely to be unlike it than like it. It follows that the desired probability is not $\frac{1}{2}$, but less than $\frac{1}{2}$. The second solution of the problem is therefore incorrect, and our faith in the first solution is restored.

*

In all the problems so far discussed the total number of possibilities – the number of ways in which the event under consideration can happen or fail to happen – has been finite. A host of contradictions arise when the number of possibilities is infinite, as is the case in the next few examples.

Given a line segment AB and any point P on AB. A point of AB is chosen at random. What is the probability that the point chosen is P?

Here the number of possibilities is evidently infinite, for it was shown in Chapter 7 that a line of finite length contains an infinite number of points. For the moment let us ignore the point P and consider a simpler problem. Suppose we divide the segment AB into 10 equal intervals, as in Figure 95. When we say that a point of AB is 'chosen at random' we mean simply that all the intervals

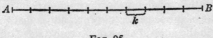

FIG. 95

are equally likely to contain the point. Hence the probability that the point chosen lies in any specified interval – say the interval labelled k – is $\frac{1}{10}$. Similarly, if AB is divided into 100 equal intervals, the probability that the random point is contained in any specified interval is $\frac{1}{100}$. And so on. Note that in every case the probability is the ratio of the length of the interval to the length of the whole segment. We can safely generalize this notion and state the following principle. *If a point is chosen at random on a line segment of length L, the probability that it falls in a specified interval of length k is k/L.*

Let us see what happens when we try to apply this principle to our original problem. For simplicity call the length of *AB* 10 inches. Now *P*, being a point, has no length. In other words, *P* occupies an 'interval' of length zero. But the probability that the random point falls in an interval of length zero is $\frac{0}{10}$, or 0. And zero probability, as we saw earlier in the chapter, means that the event cannot occur. Therefore it is impossible for the random point to coincide with *P*. Since *P* is *any* given point of *AB*, it follows that the random point cannot coincide with *any* point of the line. Hence the random point is both a point of the line and yet not a point of the line – a nice dilemma!

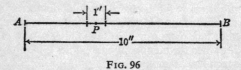

FIG. 96

To get round the difficulty here we must attack the problem from the point of view of limits, a notion we discussed in connection with infinite series. First suppose we make *P* the mid-point of an interval of length 1 inch, as in Figure 96. Then the probability that the random point lies somewhere in this interval is $\frac{1}{10}$. If we make the interval 0·1 inch long, keeping *P* the mid-point as before, the probability is $\frac{1}{100}$. If we make the interval 0·01 inch, the probability is $\frac{1}{1000}$; if 0·001 inch, $\frac{1}{10000}$; and so on indefinitely.

Now the limit of the contracting interval is the point *P* itself. Consequently the probability that the random point coincides with *P* is the limit of the sequence of probabilities that the point lies somewhere in the interval at each successive stage of its contraction. If we continue at each stage, to cut the interval down to $\frac{1}{10}$ of what it was, then the probability that the random point falls on *P* is the limit of the sequence

$$\frac{1}{10}, \frac{1}{100}, \frac{1}{1000}, \frac{1}{10000}, \frac{1}{100000}, \frac{1}{1000000}, \dots$$

The limit of this sequence, as we make the interval about *P* smaller and smaller, is zero. But this observation does not necessarily mean that the probability ever *is* zero. It simply means that we can make the probability as *near* zero as we please by making the interval about *P* sufficiently small.

At the risk of confusing the issue rather than clarifying it, let us look at a different, but more concrete, example. Suppose that a box contains 1 red marble and 9 white marbles, and that a single marble is to be drawn. Then the probability of drawing the red marble is $\frac{1}{10}$. If we increase the number of white marbles to 99, the probability of drawing the red one is $\frac{1}{100}$. If we increase the number of white marbles to 999, the probability is $\frac{1}{1000}$. And so on. As we go on adding white marbles, the probability of drawing the single red one becomes smaller and smaller, and we can make it as small as we please by adding a sufficient number of white marbles. But the probability of drawing the red marble is never zero – the red marble is always there, and there is always some chance, however small, that it will be drawn.

In a word, we must distinguish in our minds between 'zero' on the one hand and 'infinitely small', or 'infinitesimal', on the other. We can then say that although the desired probability is, for all practical purposes, zero, it is, theoretically speaking, not zero but infinitesimal. This same distinction must be made whenever the number of favourable cases is finite and the number of possible cases is infinite.

The contradictions encountered in the following two problems[7] can be handled in the manner just discussed.

Paradox 1. Since all even numbers are divisible by 2, the only even prime number is 2 itself. That is to say, the number of even primes is 1. But the total number of primes is infinite (see page 36). Therefore the probability that an arbitrary prime number is even is zero. This conclusion implies that it is impossible for a prime number to be even. Consequently the prime number 2 does not exist.

Paradox 2. The largest known prime number is $2^{127}-1$ (see p. 36). Hence the number of known primes is finite. But the total number of primes is infinite. Therefore the probability that an arbitrary prime number is known is zero. That is to say, it is impossible for a prime number to be known. Therefore no prime numbers are known.

*

The following problem[8] brings to light another difficulty which arises when the total number of cases is infinite.

A real number – rational or irrational – between 0 and 10 is chosen at random. What is the probability that it is greater than 5 ?

Using the technique we have developed in connection with a random point on a line, we divide a segment 10 units long into two intervals, each of length 5 units, as in Figure 97. Then the probability that the number chosen lies in the favourable interval is $\frac{5}{10}$, or $\frac{1}{2}$.

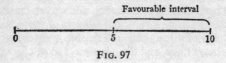

FIG. 97

Let us, for a moment, look at a related problem.

A real number – rational or irrational – between 0 and 100 is chosen at random. What is the probability that it is greater than 25 ?

This time we divide a segment 100 units long into two intervals, the first of length 25 units, the second of length 75 units, as in Figure 98. The favourable interval in this case is the second one. And the probability that the random number is greater than 25 is $\frac{75}{100}$, or $\frac{3}{4}$.

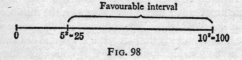

FIG. 98

Now consider the fact that every number between 0 and 25 has a square root which lies between 0 and 5, and every number between 25 and 100 has a square root which lies between 5 and 10. We can therefore interpret the results of our two problems in this way: if a *number* between 0 and 10 is chosen at random, the probability that it is greater than 5 is $\frac{1}{2}$; whereas if the *square of the number* is chosen at random, the probability that the number is greater than 5 is $\frac{3}{4}$!

What is going on here? Should not the desired probability be the same regardless of whether the number *or* its square is chosen at random? Let us scrutinize the two problems more carefully.

In the first problem we probably based our assumption that the two intervals were equally likely on the idea that the real numbers between 0 and 10 are evenly distributed along the line – that there are, so to speak, just as many real numbers between 0 and 5 as between 5 and 10. But now consider the squares of all such numbers. Every number in the interval 0 to 5 of Figure 97 has a square which lies in the interval 0 to 25 of Figure 98; and every number in the interval 5 to 10 of Figure 97 has a square which lies in the interval 25 to 100 of Figure 98. There are, in other words, just as many real numbers between 0 and 25 as between 25 and 100. (This idea is not a new one. In the last chapter we established the fact that there are as many points – corresponding to real numbers – on a line of finite length as on a line even of infinite length.) We are therefore led to the conclusion, whether we like it or not, that the intervals 0 to 25 and 25 to 100 are equally likely to contain a number picked at random between 0 and 100.

But in the second problem we went on the assumption that the numbers between 0 and 100 are distributed evenly along the line and that there are, so to speak, *three times* as many numbers between 25 and 100 as between 0 and 25. That is to say, we assumed that the interval 25 to 100 is *three times* as likely to contain the random point as the interval 0 to 25. This assumption, after all, is a reasonable one. It is the one we should have made had we had no knowledge of the first problem, but had been thinking simply of a number picked at random between 0 and 100.

The way out of all this confusion is not entirely clear. The difficulty is concerned with the proper choice of a set of equally likely cases, a matter on which the mathematicians themselves are not agreed. One group, following Bertrand, would dismiss all such problems by pointing out that infinity is not a number and that we cannot describe, in terms of finite probabilities, choices made at random from an infinitude of possibilities. This attitude indeed offers a way out, but not a very happy one, for it requires junking many results and techniques which have been found to be extremely useful.

Perhaps the most satisfactory attitude for us to take is the pragmatic one. When the number of cases is infinite, we will grant that the choice of a set of equally likely cases is arbitrary, but choose that set which common sense tells us is the most practical for the particular problem under consideration. Thus, in the two problems we have been discussing, the set used in the first problem certainly appears to be more practical for that problem than the set used in the second problem would be. What man in the street, confronted with the problem of determining the probability that a random number between 0 and 10 is greater than 5, would go off into calculations concerning the square of the random number and come out with the answer $\frac{3}{4}$? The common-sense answer is $\frac{1}{2}$.

We shall see shortly that the pragmatic attitude is not always entirely satisfactory, but the great argument in favour of it is the status of the theory of probability today. The theory is what it is because those who were responsible for its development were practical men who had the good common sense to make practical assumptions when they needed them. Had they stopped to wrangle over every theoretical point which arose, the theory might have died almost at birth. Instead, it has grown to be a powerful weapon of research in many fields.

The following paradoxes[9] illustrate how difficult it frequently is to decide what set of equally likely cases is the most workable in a given situation.

Paradox 1. Of a certain substance, it is known only that its specific volume lies between 1 and 3. It is therefore reasonable to assume that its volume is as likely to lie between 1 and 2 as between 2 and 3.

But now consider the specific density of the substance. The volume and density are related by the formula $D=1/V$. Since the volume lies between 1 and 3, the density lies between 1 and $\frac{1}{3}$. And since we know nothing else about the density, it is reasonable to assume that it is as likely to lie between 1 and $\frac{2}{3}$ as between $\frac{2}{3}$ and $\frac{1}{3}$. Consequently the volume, which is the reciprocal of the density, is as likely to lie between 1 and $\frac{3}{2}$ as between $\frac{3}{2}$ and 3; that is to say, between 1 and 1·5 as between 1·5 and 3. This conclusion, of course, contradicts our first conclusion that the volume is as likely to lie between 1 and 2 as between 2 and 3.

Paradox 2. A chord is drawn at random in a given circle. What is the probability that the chord is longer than one side of the equilateral triangle inscribed in the circle?

(a) In Figure 99, let *ABC* be the inscribed equilateral triangle, and let *DAE* be tangent to the circle at *A*. The random chord can be thought of as drawn through *A* and any other point of the circle.

172

Any chord lying within the shaded 60° angle *BAC* is longer than one side of the triangle, and is therefore a favourable case. Any chord lying within either of the 60° angles *BAD* or *CAE* is shorter

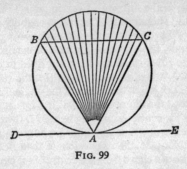

FIG. 99

than one side of the triangle. In other words, all possible cases lie within the 180° angle *DAE*, and all favourable cases within the 60° angle *BAC*. Consequently the desired probability is $\frac{60}{180}$, or $\frac{1}{3}$. The temporary fixing of the point *A* is of course no restriction, for the same argument would hold regardless of the position of *A*.

(b) Next think of the random chord as drawn perpendicular to the diameter *AK* through any point of *AK*, as in Figure 100. It is easy to show that the distance from the centre of the circle to any

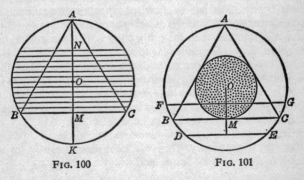

FIG. 100   FIG. 101

side of the triangle is equal to half the radius of the circle.[10] In particular, *OM* is one half the radius *OK*, or one fourth the diameter *AK*. Now it is evident that if we lay off *ON* equal to *OM*, then any

173

chord in the interval MN is greater than one side of the triangle. The random chord can be drawn through any point of AK. Chords which are longer than one side of the triangle are, as we have seen, those which lie in the interval MN – an interval whose length is half that of AK. Therefore the desired probablity is $\frac{1}{2}$. The temporary fixing of the diameter AK is no restriction, as the same argument would apply for any other position of the diameter.

(c) In Figure 101 a circle has been inscribed in the given equilateral triangle. As was pointed out in (b), the radius of the inscribed circle, OM, is half the radius of the original circle. Furthermore, a glance at the figure shows that if DE is any chord whose distance from the centre is greater than OM, then DE is shorter than BC; whereas if FG is any chord whose distance from the centre is less than OM, then FG is longer than BC. Finally, note that the distance of a chord from the centre of the circle is measured by the distance of its mid-point from the centre. Now the random chord can have as its mid-point any point within the large circle, and the mid-points of all chords having the desired property lie within the small circle. Hence the probability that the random chord is greater than one side of the equilateral triangle is the ratio of the area of the small circle to that of the large circle. If we denote the radius of the small circle by r, then the radius of the large circle is $2r$, and the ratio in question is $\pi r^2/\pi(2r)^2 = \pi r^2/4\pi r^2 = \frac{1}{4}$.

Let us summarize briefly the results of this rather lengthy example. If we assume that the chord, passing through a point on the circumference of the circle, is as likely to make one angle with the tangent as another, then the probability is $\frac{1}{3}$. If we assume that the chord, drawn perpendicular to a diameter of the circle, is as likely to pass through one point of the diameter as another, then the probability is $\frac{1}{2}$. Finally, if we assume that the mid-point of the chord is as likely to be one interior point of the circle as another, then the probability is $\frac{1}{4}$. What is the most practical set of equally likely cases here? One guess is as good as another.

(The next two paradoxes involve some knowledge of solid geometry and trigonometry. They will be discussed briefly for the benefit of those who are acquainted with these subjects.)

Paradox 3. A plane is chosen at random in space. What is the probability that it makes an acute angle of less than 45° with the plane of the horizon?

(a) The random plane can make any angle between 0° and 90° with the plane of the horizon. Only angles between 0° and 45° are favourable. Therefore the probability is $\frac{45}{90}$, or ·5.

(b) From the centre of an arbitrary hemisphere of radius r, the plane of whose base is horizontal, draw a perpendicular to the random plane. Then to choose the plane at random is to choose at random the point where the perpendicular to the plane intersects the hemisphere. If the plane is to make an angle of less than 45° with the horizontal, the perpendicular must intersect the hemisphere at some point of a zone whose area is

$$2\pi r^2(1-\cos 45°)=4\pi r^2 \sin^2 22{\cdot}5°.$$

Then the desired probability is the ratio of the area of the zone to the area of the hemisphere. This ratio is $2 \sin^2 22{\cdot}5°$, or ·293.

Paradox 4. Two points are chosen at random on the surface of a sphere. What is the probability that the distance between them is less than 10 minutes of arc?

(a) Let one of the points be fixed, and through this point draw a fixed great circle. (These restrictions are only apparent, for the argument to follow is valid for all choices of the first point and for all choices of a great circle through that point.) Now divide the great circle into 2160 equal arcs, each of length 10′. Favourable cases are those in which the second point lies in one or the other of the two arcs adjacent to the first point. Hence the desired probability is $\frac{2}{2160}$, or ·000926.

(b) The first point having been fixed, the second point can lie anywhere on the sphere. If the distance between the two points is to be less than 10′, however, the second point must lie on a zone whose area is

$$2\pi r^2(1-\cos 10′)=4\pi r^2 \sin^2 5′,$$

where r is the radius of the sphere. Therefore the desired probability is the ratio of this area to that of the sphere – that is to say, $\sin^2 5′$, or ·00000212.

This example is remarkable in that the first result is more than 400 times as large as the second!

*

The last group of paradoxes showed how difficult it is to determine the correct set of equally likely cases whenever there happens to be more than one possible set. An even more fundamental problem is that which concerns the precise meaning of 'equally likely' – a notion which is essentially intuitive and difficult to define. Indeed, the proper definition of 'equally likely cases' has split mathematicians into two opposing camps. On the one hand are the 'insufficient reasonists', who maintain that two cases are equally likely if there is no reason to think them otherwise. On the other hand are the 'cogent reasonists', who maintain that two cases are equally likely only if there is some definite reason to think them so. The distinction is in some ways a rather fine one. As a matter of fact, an insufficient reasonist might well be classed as a cogent reasonist on the ground that, to him, the most cogent reason for thinking two things equally likely is the absence of any reason for thinking them otherwise![11]

We who have been following the discussions of this chapter should probably be classed as insufficient reasonists because of the fact that we have spent very little time looking for good reasons why we should assume that two or more cases are equally likely. Consider, for example, the first of the last four paradoxes discussed – the one concerning volume and density. We were told that we knew nothing of a certain substance other than that its volume had some value between 1 and 3. In the absence of all other information, we assumed that the volume was as likely to lie between 1 and 2 as between 2 and 3. The cogent reasonist would never have let himself in for the difficulties we encountered in this problem, for at the very start he would have dismissed the problem as one which simply cannot be discussed.

The classic example used by the cogent reasonist to confound the insufficient reasonist is the so-called 'life on Mars paradox'. We shall present this paradox in the form of a dialogue between the cogent reasonist (C. R.) and the insufficient reasonist (I. R.).

C. R.: Tell me, Mr I. R., what in your opinion is the probability of life, in some form or other, on the planet Mars?

I. R.: H'm, let me see. Well, since I am totally ignorant of the answer, I shall have to assume that the possibilities of life and no life are equally likely. Therefore my answer is $\frac{1}{2}$.

C. R.: Very good. But now let us look at the problem from another angle. What would you say is the probability of no horses on Mars?

I. R.: Again I confess total ignorance, so again, I must conclude $\frac{1}{2}$.

C. R.: And the probability of no cows?

I. R.: Again $\frac{1}{2}$

C. R.: And the probability of no dogs?

I. R.: Again $\frac{1}{2}$.

This sort of thing goes on for several minutes, while C. R. names, let us say, 17 more specific forms of life.]

C. R.: Very well. But now we must conclude that the probability of all these things occurring at once – no horses *and* no cows *and* no dogs *and* none of the other 17 forms of life which I specified – is the product of the individual probabilities, or $(\frac{1}{2})$ $(\frac{1}{2})$ $(\frac{1}{2})$ ... to twenty terms. *If the reader has forgotten the principle involved here, he should refer back to page 163.*] In other words, the probability that none of these twenty forms of life exists is $(\frac{1}{2})^{20}$, or $\frac{1}{1048576}$. Am I right so far?

I. R. (*beginning to understand the trouble for which he is heading*): Why, yes, I am afraid you are.

C. R.: Thank you. But if the probability that *none* of these forms of life exists is $\frac{1}{1048576}$, what, may I ask, is the probability that *at least one* of them exists?

I. R.: Unfortunately, I must confess that this probability is the difference between your result and 1 – that is to say, $\frac{1048575}{1048576}$.

C. R.: And so, Mr I. R., we are led to two results concerning the probability of life on Mars. One of these is ·5, and the other is about ·999999 – very near to certainty. Surely one of the two must be wrong. Can it be that your principle of insufficient reason is at fault?

Poor Mr I. R.! We have, of course, presented the dialogue as C. R. might have written it. Perhaps we can find something to say in defence of I. R. Note that the paradox is based on two assumptions. In both solutions it is necessary to assume that we have absolutely no information concerning the existence or non-existence of life on Mars. And in the second solution it is necessary to assume that the occurrence of one form of life is absolutely independent of the occurrence of any other form of life – otherwise the multiplicative principle used in the argument would not apply. It is true that both these assumptions might be valid in a purely hypothetical universe, but the knowledge we have of our own universe makes them ridiculous. Once again the question is one of practicality. We *do* know something of the planet Mars, and we *do* know something of the dependence of one form of life on another. These two facts are sufficient to invalidate the argument of the cogent reasonist.

Much the same sort of difficulty is involved in a problem, discussed by Bertrand, concerning weather predicions.[12]

Suppose one forecaster predicts that it will be fair tomorrow, and the probability that he is wrong is $\frac{1}{5}$. Suppose a second forecaster predicts the same, and the probability that he is wrong is also $\frac{1}{5}$. Then the probability that both are wrong would appear to be $(\frac{1}{5})(\frac{1}{5})$, or $\frac{1}{25}$.

But are the two predictions independent? Suppose the two forecasters have been educated at the same school, that they have adopted the same principles and that they base their predictions on the same data. Then if one is wrong, the other will be wrong also, and the second factor of the above product is not $\frac{1}{5}$, but 1. In other words, the accord of the two predictions does not lessen the chance of error.

To cap the argument, suppose that one predicts 'rain' and the other 'clear'. Assuming that 'rain' means 'rain all day long' and that 'clear' means 'clear all day long', the probability that they are both right is not $(\frac{4}{5})(\frac{4}{5})$, but, since the occurrence of this event is impossible, zero.

Perhaps news of this sort should not be spread about. Wealthy people afflicted with interesting maladies may, if they hear of it, have less confidence than usual in the coincident opinions of their three or four expensive specialists.

*

One of the most famous of all probability paradoxes is, like the problem which opened this chapter, a gambling problem. This is the 'St Petersburg paradox', originally proposed by Nicolaus Bernoulli in a letter dated September 1713. The original problem was modified by Daniel Bernoulli – nephew of Nicolaus – and discussed at length by him in the *Transactions of the St Petersburg Academy*. Here it received its notoriety and its name. (It may be worth remarking, in passing, that the Bernoulli family produced eight mathematicians in three generations!)

178

A coin is tossed until heads appears. If heads appears on the first toss, the bank pays the player £1. If heads appears for the first time on the second toss, the bank pays £2. If heads appears for the first time on the third toss, £4; on the fourth toss, £8; on the fifth toss, £16; and so on. What amount should the player pay the bank for the privilege of playing one game in order that the game be fair – that is to say, in order that neither the player nor the bank has an advantage regardless of how long the game goes on?

First let us be sure we know what is meant by a 'fair game'. Consider the following simple example. A player undertakes to throw a four with one throw of a single die. The bank agrees to pay him £1 if he succeeds. What about should the player pay if the game is to be a fair one?

In a single throw the probability of a four is obviously $\frac{1}{6}$. Now we cannot infer from this that the player will throw 1 four in exactly 6 throws. We can infer, however, that in a large number of throws – say 6000 – a four will occur *about* 1000 times, and that as we increase the number of throws, the ratio of the number of successes to the number of throws will approach more and more nearly to $\frac{1}{6}$. (This is an application of a theorem enunciated by Jacob Bernoulli, brother of Nicolaus.) The player's 'expectation', as it is called, is therefore $\frac{1}{6}$ of £1 per game, and this amount is what he should pay the bank if neither he nor the bank is to have an advantage.

One more example. Suppose the bank agrees to pay the player £1 if he gets a four on the first throw. If he fails, it will pay him £1 if he gets a four on the second throw. What amount should the player pay the bank in this instance? As before, the player's expectation on the first throw is $\frac{1}{6}$ of £1. His expectation on the second throw, however, is not $\frac{1}{6}$ of £1. He will collect on this throw only if he fails to collect on the first throw. Now the probability that he does *not* get a four on the first throw is $\frac{5}{6}$, and the probability that he *does* get a four on the second throw is $\frac{1}{6}$. Hence the probability that he fails on the first and succeeds on the second is $(\frac{5}{6})(\frac{1}{6})$, or $\frac{5}{36}$. That is to say, his expectation on this throw is $\frac{5}{36}$ of £1. Finally, the player's chance of collecting on the first throw *or* the second throw is $\frac{1}{6} + \frac{5}{36}$, or $\frac{11}{36}$. His expectation in this game is therefore $\frac{11}{36}$ of £1 – the amount he should pay if the game is to be fair. Note that here, as in any game of this kind, the total expectation is the sum of the expectations at each stage of the game.

And now back to the original problem. Consider the first toss of the coin. The probability of heads is $\frac{1}{2}$. The amount involved is £1. Therefore the expectation on this toss is $\frac{1}{2}$ of £1, or 10s. Consider the second toss. The player will collect on this toss only if he throws tails on the first toss and heads on the second. The probability that this will happen is $(\frac{1}{2})(\frac{1}{2})$, or $\frac{1}{4}$. The amount involved is £2. Therefore the expectation on this toss is $\frac{1}{4}$ of £2, or 10s. Consider the third toss. The player will collect on this toss only if he throws tails on the first two tosses and heads on the third. The probability that this will happen is $(\frac{1}{2})(\frac{1}{2})(\frac{1}{2})$, or $\frac{1}{8}$. The amount involved is £4. Therefore the expectation on this toss is $\frac{1}{8}$ of £4, or 10s.

To show that the expectation on *every* toss is 10s, consider the nth toss. The player will collect on this toss only if he throws tails on the first $n-1$ tosses and heads on the nth. The probability that this event will happen is $(\frac{1}{2})(\frac{1}{2})(\frac{1}{2}) \ldots (\frac{1}{2})$ to n factors, or $\frac{1}{2^n}$. Now the number of pounds involved in the first toss is 1, or 2^0; that in the second toss, 2, or 2^1; that in the third toss, 4, or 2^2; that in the fourth toss 8, or 2^3; and so on. Note that the number of pounds is always a power of 2, and that the power is always one less than the number of the toss. Hence the number of pounds involved in the nth toss is 2^{n-1}. Finally, then, the expectation on the nth toss is $(\frac{1}{2^n})(2^{n-1})$, or $2^{n-1}/2^n$, or 10s.

Since the total expectation is always the sum of the expectations at each stage of the fame, the total expectation here is

$$£\frac{1}{2}+\frac{1}{2}+\frac{1}{2}+\frac{1}{2}+\frac{1}{2}+\frac{1}{2}+\ldots$$

Now recall that play is to continue until heads turns up. Theoretically there is no limit to the number of tails which may appear before the first head appears, and this means that the above series is to be summed to infinity. But the sum of an infinite number of terms of this series is obviously infinite. It follows that *the player must pay the bank an infinite amount of money for the privilege of playing one game!*

This result is absurd. No one would ever think of paying any great amount for such an opportunity. Yet the mathematics is correct. What is wrong, then? This question has been bothering mathematicians for some two hundred years, and as yet no one has found an answer acceptable to all concerned. A number of solutions have been suggested.[13] Of these, the following one probably appeals most to common sense.

There is nothing wrong with the result we arrived at provided there exists a bank which has infinite wealth and which is consequently in a position to pay the player no matter how late in the game the first head turns up. But such a bank obviously does not exist. Suppose, then, that we investigate the expectation in the case of a bank whose wealth is limited to £1,000,000.

As before, the probability that a head first appears on the nth toss is $\frac{1}{2}^n$. If a head does appear on this toss, the bank pays £2^{n-1} provided this amount is less than £1,000,000. Otherwise it pays £1,000,000. That is to say, if p_n denotes the probability that a head first appears on the nth toss, and if a_n is the amount in pounds paid by the bank for a win on that toss, then the expectation on the nth toss is $p_n.a_n$, where

$$\left.\begin{array}{l} p_n = \dfrac{1}{2^n} \\[2mm] a_n = 2^{n-1} \end{array}\right\} \text{provided } 2^{n-1} \text{ is less than } 1,000,000,$$

$$\left.\begin{array}{l} p_n = \dfrac{1}{2^n} \\[2mm] a_n = 1,000,000 \end{array}\right\} \text{provided } 2^{n-1} \text{ is greater than } 1,000,000.$$

Now 2^{19} is less than 1,000,000, while 2^{20} is greater than 1,000,000. It follows that the first set of conditions applies when n is less than or equal to 20, and the second set when n is greater than 20. Therefore the total expectation in pounds is given by the expression

$$\frac{1}{2}(1) + \frac{1}{2^2}(2) + \frac{1}{2^3}(2^2) + \frac{1}{2^4}(2^3) + \dots \text{ to twenty terms}$$

$$+ \frac{1}{2^{21}}(1,000,000) + \frac{1}{2^{22}}(1,000,000) + \dots \text{ to infinity.}$$

Since each of the first twenty terms of this series has the value $\frac{1}{2}$, the sum of the first part of the series is 10. The second part is a geometric series, the sum of which can be obtained by an elementary algebraic formula. The value of this sum to four decimal places is ·9536. The total expectation in the case of a £1,000,000 bank is thus seen to be £10·95, a not unreasonable amount to pay for the privilege of playing.

While we are on the subject of gambling, here are two hints on how to win at roulette. They are included for the benefit of those who wish to take them for what they are worth. The author assumes no responsibility in connection with either of them!

Paradox 1. If you are willing to take a chance on losing £10 – but no more than £10 – proceed as follows. Put £10 on red (or black) the first day. If you win, put £20 on red the second day. If you win, put £30 on red the third day. Continue as long as you win. If you lose, stop at once and never play again. Then if you ever lose, you lose no more than £10. But if you continue to win, you will win $10+20+30+...+10n$ pounds by stopping after the 1st, 2nd, 3rd, ..., nth day.

Paradox 2. If you wish always to be ahead of the bank, stick to one wheel and play consecutive games as follows:[14] Bet £1 on red. If you win, all well and good. If you lose, put £2 on red. If you then win, you are £1 ahead. If you lose, put £4 on red. If you then win, you are again £1 ahead. If you lose, put £8 on red. And so on. Keep betting until you win. Theoretically, of course, it is possible for the bank to wipe you out financially. Actually, however, runs of more than 10 or 12 successive blacks or red are extremely rare, and your stake at the twelfth play would be only £2048. When you *do* win you will, as before, be £1 ahead of the bank. You can then begin all over again. Simple, isn't it?

*

As a final example of the pitfalls of probability we shall consider an amusing paradox proposed by Lewis Carroll.[15]

A bag contains two counters as to which nothing is known save that each is either black or white. Ascertain their colours without taking them out of the bag.

Carroll insisted that the answer is 'one white, one black' by the following argument: We know that if a bag contains three counters, two being black and one white, the probability of drawing a black one is $\frac{2}{3}$, and no other state of things will give this probability.

Now with two counters there are four equally likely cases: both counters can be black, or the first black and the second white, or the first white and the second black, or both white. For brevity we shall denote these cases by *BB*, *BW*, *WB*, and *WW* respectively.

Since they are equally likely, and since one of them must represent the true situation, the probability of each is $\frac{1}{4}$.

Add a black counter, Then, as before, the probabilities of *BBB*, *BWB*, *WBB*, and *WWB* are each $\frac{1}{4}$. Now note that in the case of *BBB*, the probability of drawing a black counter is 1; in that of *BWB*, $\frac{2}{3}$; in that of *WBB*, $\frac{2}{3}$; and in that of *WWB*, $\frac{1}{3}$. Therefore the probability of drawing a black counter from the bag is

$$1 \cdot \frac{4}{1} + \frac{2}{3} \cdot \frac{1}{4} + \frac{2}{3} \cdot \frac{1}{4} + \frac{1}{3} \cdot \frac{1}{4} = \frac{3}{12} + \frac{2}{12} + \frac{2}{12} + \frac{1}{12} = \frac{8}{12}, \text{ or} \frac{2}{3}.$$

But, as we have said before, the probability of drawing a black counter is $\frac{2}{3}$ only if the bag contains two black counters and one white counter. Hence, before the black counter was added, the bag must have contained one white, one black!

It does not take long to see that the adding of the black counter is little more than rigmarole designed to confuse the reader. For suppose we return to the original situation involving only two counters. The possible cases are

$$BB, BW, WB, WW;$$

the probabilities of these cases are

$$\frac{1}{4}, \frac{1}{4}, \frac{1}{4}, \frac{1}{4};$$

and the probabilities of drawing a black counter in the respective cases are

$$1, \frac{1}{2}, \frac{1}{2}, 0.$$

By the same argument that was used before, the probability of drawing a black counter is, in the combined cases,

$$1 \cdot \frac{1}{4} + \frac{1}{2} \cdot \frac{1}{4} + \frac{1}{2} \cdot \frac{1}{4} + 0 \cdot \frac{1}{4} = \frac{2}{8} + \frac{1}{8} + \frac{1}{8} + 0 = \frac{4}{8}, \text{ or } \frac{1}{2}.$$

But if the probability of drawing a black counter is $\frac{1}{2}$, and if there are two counters in the bag, one must be white and the other black.

The paradoxical conclusion does not, therefore, depend upon the adding of a third counter. The fallacy lies in the third step – that in which the probabilities of drawing a black counter in the individual cases are combined to give a single probability. Perhaps the easiest way to convince ourselves of this fact is to carry through the argument for a bag containing *three* counters.

If there are three counters, each of which can be either black or white, the possible cases are

BBB, BBW, BWB, WBB, BWW, WBW, WWB, WWW.

Since there are eight equally likely cases, one of which must represent the true state of things, the probability of each is

$$\frac{1}{8}, \frac{1}{8}, \frac{1}{8}, \frac{1}{8}, \frac{1}{8}, \frac{1}{8}, \frac{1}{8}, \frac{1}{8}.$$

The probabilities of drawing a black counter in these cases are, respectively,

$$1, \frac{2}{3}, \frac{2}{3}, \frac{2}{3}, \frac{1}{3}, \frac{1}{3}, \frac{1}{3}, 0.$$

If these probabilities are combined as before, the probability of drawing a black counter is

$$1 \cdot \frac{1}{8} + \frac{2}{3} \cdot \frac{1}{8} + \frac{2}{3} \cdot \frac{1}{8} + \frac{2}{3} \cdot \frac{1}{8} + \frac{1}{3} \cdot \frac{1}{8} + \frac{1}{3} \cdot \frac{1}{8} + \frac{1}{3} \cdot \frac{1}{8} + 0 \cdot \frac{1}{8}$$

$$= \frac{3}{24} + \frac{2}{24} + \frac{2}{24} + \frac{2}{24} + \frac{1}{24} + \frac{1}{24} + \frac{1}{24} + 0$$

$$= \frac{12}{24}, \text{ or } \frac{1}{2}.$$

But if the probability of drawing a black counter is $\frac{1}{2}$, the number of black counters must be equal to the number of white counters – a situation which simply cannot exist in the case of three counters. The same argument applied to any number of counters will always give the same result, $\frac{1}{2}$. Consequently the argument is not a valid one.

*

At least twice in this chapter remarks were made concerning the applicability of the theory of probability to other fields. A discussion of the role played by probability in the theoretical sciences – notably physics and chemistry – would lead us too far astray into technical matters. But a few words about its relation to the applied sciences may help to give us some idea of its importance in everyday activities.

In economics, for example, statistical methods – and statistics and probability are inseparable – have been found to be indispensable in the study of insurance, benefit and pension plans, market surveys, and demand and price fluctuations. Industry uses statistics extensively in such matters as the inspection of items manufactured in mass production, and the subsequent improvement of manufacturing processes. And even those engaged in modern warfare are finding probability and statistics helpful in their attempts to increase the accuracy and effectiveness of their gunnery and bombing.

Early in the eighteen hundreds Laplace – who was not only a mathematician, but also one of France's greatest astronomers and physicists – hailed the theory of probability as 'the most important object of human knowledge'. This estimate may have seemed reckless at the time it was made, but in these days it is beginning to seem somewhat more sound.

CHAPTER NINE

Paradoxes in Logic

'MATHEMATICS and logic, historically speaking, have been entirely distinct studies. Mathematics has been connected with science, logic with Greek. But both have developed in modern times: logic has become more mathematical and mathematics has become more logical. The consequence is that it has now become wholly impossible to draw a line between the two; in fact, the two are one.... The proof of their identity is, of course, a matter of detail: starting with premises which would be universally admitted to belong to logic, and arriving by deduction at results which as obviously belong to mathematics, we find that there is no point at which a sharp line can be drawn, with logic to the left and mathematics to the right.'

So wrote Bertrand Russell in 1919.[1] In spite of the fact that many mathematicians still refuse to admit the *identity* of mathematics and logic, there is, as Russell indicates, ample evidence of the fact that a close relationship does exist between the two subjects. Anyone fortunate enough to have studied plane geometry under the right kind of teacher is at least mildly aware of this relationship, although such matters are sadly neglected in most elementary courses. Certainly the connection between mathematics and logic is close enough for contradictions in logic to have a disquieting effect on mathematics.

The troublesome issues raised by the paradoxes of logic can for the most part be traced to one basic cause. Furthermore the paradoxes are, if not amusing, at least thought-provoking in themselves. We shall consequently examine them first with little regard for their mathematical significance, and shall then return to scrutinize them more carefully.

*

The oldest of the logical paradoxes was discussed, in simplified form, at the very beginning of this book. It dates back to the sixth century B.C. when Epimenides, the celebrated poet and prophet of Crete, is supposed to have made his famous remark, 'All Cretans are liars.' If we are to find anything paradoxical in this remark, we must rewrite it in the form, 'All statements made by Cretans are false.'

Now offhand this does not appear to be a particularly dangerous verdict. It resembles the idle exaggerations in which all of us indulge – such things as 'All the stars are out tonight,' 'All the books that have appeared this season are worthless,' and 'All the shop-keepers in this town are thieves.' But 'All statements made by Cretans are false' is much more than an idle exaggeration. Like the fabulous hoop snake, it suddenly turns and starts swallowing itself. The trouble begins when we consider the fact that Epimenides, who makes this statement, is himself a Cretan. In that case, all statements made by Epimenides are false. In particular, his statement 'All statements made by Cretans are false' is false, so that all statements made by Cretans are *not* false.

We are now probably so bogged down in words that we don't know *where* we are. That, unfortunately, is one of the difficulties we encounter in all these paradoxes. Their significance is seldom apparent at first reading – they must be read and re-read until they are clear. It will perhaps be of some help here to put the argument in a step-by-step form.

(1) All statements made by Cretans are false.
(2) Statement (1) was made by a Cretan.
(3) Therefore statement (1) is false.
(4) Therefore all statements made by Cretans are not false.

Now statements (1) and (4) obviously cannot both be true, yet statement (4) follows logically from statement (1). Consequently statement (1) is self-contradictory.

*

There is hardly a person living who has not made use, at some time or other, of the well-worn adage, 'All rules have exceptions.' There are probably few people, however, who are aware of the fact that it is self-contradictory.

The statement is, to all intents and purposes, a rule to the effect that all rules whatever have their exceptions. Now if *all* rules have exceptions, then this particular rule – 'All rules have exceptions' – must have an exception. And what would an exception to this rule be? Why, the only thing it could be is a rule *without* an exception. And if a rule without an exception exists, then all rules do not have exceptions.

But perhaps we had better resort once more to the step-by-step argument.

(1) All rules have exceptions.
(2) Statement (1) is a rule.
(3) Therefore statement (1) has exceptions.
(4) Therefore all rules do not have exceptions.

*

Another paradox which has its foundation – real or legendary – in antiquity concerns the sophist Protagoras, who lived and taught in the fifth century B.C. It is said that Protagoras made an arrangement with one of his pupils whereby the pupil was to pay for his instruction after he had won his first case. The young man completed his course, hung up the traditional shingle, and waited for clients. None appeared. Protagoras grew impatient and decided to sue his former pupil for the amount owed him.

'For,' argued Protagoras, 'either I win this suit, or you win it. If I win, you pay me according to the judgement of the court. If you win, you pay me according to our agreement. In either case I am bound to be paid.'

'Not so,' replied the young man. 'If I win, then by the judgement of the court I need not pay you. If you win, then by our agreement I need not pay you. In either case I am bound not to have to pay you.'

Whose argument was right? Who knows?

*

A stranger in town once asked the barber if he had much competition. 'None at all,' replied the barber. 'Of all the men in the village, I naturally don't shave any of those who shave themselves, but I do shave all those who don't shave themselves.'

This remark appears innocent enough until we stop to think of the plight of the barber. Does he shave himself or doesn't he? Let's suppose he *does*. Then he is to be classed with those who shave themselves. But the barber doesn't shave those who shave themselves. Therefore he does *not* shave himself. All right, then, let's suppose he does *not* shave himself. Then he is to be classed with those who don't shave themselves. But the barber shaves all those who don't shave themselves. Therefore he *does* shave himself.

Here is an intolerable situation. For if the poor barber shaves himself, then he doesn't, and if he doesn't, he does. Even growing a beard won't help him!

*

If we care to do so we can express every integer in simple English, without the use of numerical symbols. For example, 7 can be expressed as 'seven', or as 'the seventh integer', or as 'the third odd prime'. Again 63 can be expressed as 'sixty-three', or as 'seven times nine'. And 7396 can be expressed as 'seven thousand three hundred and ninety-six', or as 'seventy-three hundred and ninety-six', or as 'eighty-six squared'.

It is at once evident that to express each integer requires the use of a certain number of syllables. In general, the larger the number, the more syllables required. This generalization is not always true, however. For example, the thirty-nine digit number on page 36 can be expressed in five syllables as 'the largest known prime'. The important thing to note is that every integer requires a certain *minimum* number of syllables.

Now let us divide all integers into two groups, the first to include all those which require a minimum of eighteen syllables or less, the second to include all those which require a minimum of nineteen syllables or more. Consider the second group. Of all the members of this group, there certainly is one which is the smallest. Just what integer constitutes this smallest member is beside the point. It is sufficient to note that 'the least integer not namable in fewer than nineteen syllables' is *some specific number*.

But what of the phrase in quotation marks? It is certainly one way of expressing, in English, the smallest member of the second group. And this phrase requires only *eighteen* syllables – count them. In other words, the least integer not namable in fewer than nineteen syllables can be named in eighteen syllables!

*

Consider next all the adjectives in the English language. Each adjective has a certain meaning. In some adjectives the meaning applies to the adjective itself; in others it does not. For example, 'short' is a short word, but 'long' is not a long word. 'English' is an English word, but 'French' is not a French word. 'Single' is a single word, but 'hyphenated' is not a hyphenated word. 'Polysyllabic' is a polysyllabic word, but 'monosyllabic' is not a monosyllabic word. And so on.

Since the meaning of an adjective must either apply to itself or not apply to itself, we can divide all adjectives into two groups accordingly. Thus, if the meaning of a given adjective applies to itself, we shall classify it as 'autological'. And if its meaning does not apply to itself, we shall classify it as 'heterological'.

Now let us consider the word 'heterological'. This word is certainly an adjective, and so it must be either autological or heterological. But if 'heterological' is heterological, then this very statement – *'heterological' is heterological* – asserts that 'heterological' applies to itself. And if it does apply to itself, then it must be autological, according to our definition of the word 'autological'. On the other hand, if 'heterological' is autological, then this very statement – *'heterological' is autological* – asserts that 'heterological', being autological and not heterological, does not apply to itself. And if it does not apply to itself, then it must be heterological, according to our definition of the word 'heterological'.

The situation that confronts us is appalling. A given adjective must obviously be either autological *or* heterological – it cannot be both autological *and* heterological. Yet we have just shown that if the adjective 'heterological' is heterological, it is *not* heterological, but autological; and if it is autological, it is *not* autological, but heterological!

*

But enough, for the moment, of examples. It is time that we stopped to think about the nature of the difficulties involved in them, and to see in what way they affect mathematics.

Note first that there is one characteristic common to all these paradoxes. *They are concerned with statements about 'all' the members of certain classes of things, and either the statements or the things to which the statements refer are themselves members of those classes.* This common characteristic is not as obvious in some instances as in others. For this reason it will be well to review our examples briefly with an eye to recognizing the characteristic in each of them.

'All statements made by Cretans are false.' Since this is a statement made by a Cretan, it is itself a member of the class of all statements made by Cretans. Here the characteristic is obvious, as it is in the case of 'All rules have exceptions.'

The problem of Protagoras and his pupil is concerned with the class of all cases to be argued in court by the pupil. Included in this class is the case built round the class itself.

The quandary of the village barber is concerned with the class of all men in the village who either shave themselves or do not shave themselves. Since the barber either shaves himself or does not shave himself, he is evidently a member of this class.

The 'least integer' problem involves the class of all English expressions denoting integers. The phrase 'the least integer not namable in fewer than nineteen syllables' is an English expression denoting an integer, and so is a member of the class in question.

Finally, in the last paradox discussed, the class of all adjectives, autological or heterological, obviously includes the adjective 'heterological'.

The vicious circle which arises when a statement is made about 'all' the members of a certain class, and when the statement or the thing to which the statement refers is itself a member of that class, is difficult to avoid. Bertrand Russell, as early as 1906, tried to get round the difficulty by means of what he called the 'theory of logical types'.[2] He held that logical entities – statements, rules, things, and the like – are not all of one type, but fall into a hierarchy of types which are radically different, however similar they may appear to be. Moreover, whatever involves 'all' of a certain class of things is not of the same type as the things themselves. Take the case of 'All statements made by Cretans are false.' The 'statements referred to are statements about things. The statement *itself* is not a statement about things, but a *statement about statements* about things. It is therefore a statement of a different type, and so cannot be made to refer to itself. Hence it can lead to no contradiction. Similarly, in the rule, 'All rules have exceptions,' the 'rules' referred to are rules about things, whereas the rule *itself* is not a rule about things, but a *rule about rules* about things.

The theory of types, as described above, appears to be a safe and fairly simple means of escape from the troublesome vicious circles. Actually, however, the difficulties involved in the paradoxes are much more subtle than we have made them seem.[3] But rather than get any deeper into a discussion of the efficacy of the theory of types, let us stop to consider the following question – one which many of us have been waiting patiently to have answered. *What have the logical paradoxes to do with mathematics?* We shall try to answer this question by means of three more paradoxes. They are different from those we have discussed in that they bear directly on mathematics, yet they are the same in that the contradictions involved arise from what is essentially the same basic source.

*

The first of our three paradoxes has to do with transfinite numbers, a subject we discussed in the last section of Chapter 7. Recall that we examined two transfinite numbers in detail: A_1, the number of natural numbers, and C, the number of real numbers. Recall further our remarking that Cantor proved conclusively that just as there is no greatest natural number, so there is no greatest transfinite number. His proof hinges, essentially, on one of the properties of transfinite numbers noted on page 158. That is to say, the number 2, raised to any transfinite power, always generates a new – and larger – transfinite number. Thus $2^{A_1} = C$, $2^C = $ a still larger transfinite, and so on.

But now consider the *class of all classes*. And we mean ALL classes – all books, all chairs, all plants, all animals, all numbers (finite or transfinite, real or imaginary, rational or irrational), all things which ever existed in this or any other universe, all ideas you or anyone else, living or dead, may ever have had – everything conceivable goes into this class. Now surely no class can ever have more members than this, the class of all classes. But if such is the case, then the transfinite number of this class is unquestionably the greatest transfinite number. Yet, as we have said, Cantor proved that there is no such thing as a greatest transfinite number!

This paradox was brought to light by the Italian mathematician, Burali-Forti, in 1897. As originally conceived and stated,[4] it involves a number of technical terms and ideas which for lack of space in Chapter 7 we neglected to develop. The non-technical description given above is consequently by no means as accurate as it should be. The whole thing may seem to us to be a case of rather fine hair-splitting, but its importance to mathematics is indicated by the fact that when it first appeared it nearly brought about the collapse of the entire Cantorian theory.

*

The difficulties involved in our second paradox are similar to those we encountered in the autological–heterological controversy.

Note first that classes are either members of themselves or not.

For example, the class of all entities is itself an entity, while the class of all men is not a man. The class of all ideas is itself an idea, while the class of all stars is not a star. The class of all classes – the class we worked with in the last paradox – is itself a class, while the class of all books is not a book. And so on.

Since any given class must be either a member of itself or not a member of itself, we can divide all classes into two groups accordingly. We shall denote by *S* the class of all self-membered classes – that is, classes which are members of themselves. And we shall denote by *N* the class of all non-self-membered classes – that is, classes which are not members of themselves.

Now let us fix our attention on *N*. Since *N* is a class, it must be either self-membered or not. That is to say, *N* must be a member either of *S*, the class of all self-membered classes, or a member of *N*, the class of non-self-membered classes. If *N* is a member of *N*, then this very statement – *N is a member of N* – asserts that *N* is a member of itself. And if *N* is a member of itself, it must be a member of *S*, the class of all self-membered classes. On the other hand, if *N* is a member of *S*, then this very statement – *N is a member of S* – asserts that *N*, being a member of *S*, is not a member of itself, or *N*. And if *N* is not a member of itself, it must be a member of *N*, the class of all non-self-membered classes.

Now obviously a given class must be either self-membered or non-self-membered – it cannot be both. In other words, a given class must be a member of either *S or N* – it cannot be a member of both *S* and *N*. Yet we have just shown that if the class *N* is a member of *N*, it is *not* a member of *N*, but of *S*; and if *N* is a member of *S*, it is *not* a member of *S*, but of *N*!

The same argument is usually presented by mathematicians and logicians in a much more compact form. This form may appeal to those of us who were confused by the wordiness of the argument as presented above. Let's try it.

Denote any class by *X*, and, as before, the class of all non-self-membered classes by *N*. Then the following statement is true.

X is a member of N if and only if X is not a member of X. That is to say, *X* is a member of the class of all non-self-membered classes if and only if *X* is not a member of itself. Since *X* represents *any* class, and since *N* is a class, we may substitute *N* for *X*. The statement then reads

N is a member of N if and only if N is not a member of N.

Here again is what may seem to us to be a hair-splitting proposition as far as mathematics is concerned. Its significance, however, is made apparent by the following historical note. Gottlob Frege, a German mathematician, had spent years in an attempt to put mathematics on a sound logical basis. His chief work was a two-volume treatise on the foundations of arithmetic, a treatise in which he used freely the notion of a class of all classes that have a given property. Some indication of the time he spent on this work is to be had from the fact that the first volume was published in 1893, the second in 1903. As the second volume was about to appear, Bertrand Russell sent Frege the paradox we have just discussed. Frege acknowledged the communication as follows at the end of his second volume.

'A scientist can hardly meet with anything more undesirable than to have the foundation give way just as the work is finished. In this position I was put by a letter from Mr Bertrand Russell as the work was nearly through the press.'

Incidentally, Frege's use of the word 'undesirable' makes his remark one of the great understatements of all time!

*

The third of the three paradoxes we set out to discuss is the so-called Richard paradox,[5] named after its originator, J. Richard, a French mathematician.

Before getting into the main argument, let us consider a simple analogy. Suppose that our English vocabulary consisted of only three words – say 'see', 'the', and 'cat'. Now it stands to reason that with such limitations we could never discuss any idea which requires *more* than three words. We could not, for example, develop any of the ideas we now have under consideration. This conclusion may appear to be childishly simple, but it will clarify the argument to follow.

Any system of symbolic logic, or mathematics, consists of a collection of formulas. Here we are using the word 'formula' not in the restricted mathematical sense, but in the broadest sense. That is to say, a formula is any symbol (including letters of the alphabet, numbers, punctuation marks, and the like), or any word, or definition, or statement, or theorem – anything by which we express ideas. Now it is not difficult to show, using the notion of one-to-one correspondences which we developed in connection with transfinite numbers, that it is possible to set up a one-to-one correspondence between the class of all formulas of any given system and the class of all natural numbers. In other words, the class of all formulas has the transfinite number A_1.

The Richard paradox then consists in what amounts to the following problem: How can any system of symbolic logic, in which the class of all formulas has the transfinite number A_1, be adequate for the discussion and development of any branch of mathematics that deals with classes whose transfinite numbers are greater than A_1? In particular, how can we even talk about the class of real numbers, whose transfinite number, C, has been proved to be greater than A_1?

*

It must be emphasized again that the paradoxes of logic are not foolish problems with which the philosophically-minded while away their time. It is true that they may have existed as such for hundreds of years. But when, at the beginning of the present century, Burali-Forti, Russell, and Richard dressed them up and paraded them in mathematical costumes, they started a revolution which is still very much in progress. It is almost impossible to discuss – briefly and in non-technical terms – what is going on in this revolution. But enough can be said to give us at least some idea of present trends.[6]

Those who are engaged in laying the new foundations of mathematics can be roughly divided into the following three groups: (1) the logistic group, led by the Englishman, Bertrand Russell; (2) the axiomatic group, led by the German, David Hilbert; (3) the intuitionist group, led by the Dutchman, L. E. J. Brouwer.

The programme of the logistic group is to reduce mathematics to symbolic logic. This fact might have been inferred from the passage quoted at the beginning of the present chapter. As we have already seen, Russell proposed the theory of types as a means of getting round the logical contradictions. The shortcomings of this particular technique have been recognized, and repeated attempts have been made to modify it accordingly. It is still not entirely satisfactory.

The programme of the axiomatic group is to base all mathematics on a fundamental system of axioms, or assumptions. Such systems have been found for important parts of mathematics, and it remains only to prove that the systems are consistent – that no contradictions can arise in results deduced from them. In the case of many of these systems it has been shown that any contradiction arising from the system would imply a contradiction in arithmetic. Thus the chief problem of the axiomatic group is that of proving that the axioms of arithmetic are consistent. No satisfactory proof of this has yet been found.

Finally, the intuitionist group maintains that no mathematical concept is admissible unless it can be constructed. That is to say, not only must the concept exist in name, but an actual construction must be exhibited for the thing which the concept represents. Now if the construction is to be an actual one, then it must consist of a finite number of steps – or, as the Richard paradox indicates, of certainly no more than A_1 steps. And in that case we have no right to talk about such a thing as the class of all real numbers, whose transfinite number is greater than A_1. This attitude is hardly satisfactory, for it means that many of the most powerful and useful methods of mathematics must be thrown overboard.

Which of these three groups has the 'best' policy? The answer is a matter of opinion. As in politics, each individual interested in the controversial issues must ally himself with that party whose platform seems to him the most reasonable. However dissimilar their paths may be, all three groups are working toward the same end – to establish all mathematics on an unassailably sound basis. No one can predict whether or not this ideal will ever be attained. But already the controversies of the last few decades have brought forth entirely new fields of research, as well as new and effective methods in old fields. So as far as mathematics as a whole is concerned, the setbacks occasioned by the paradoxes of logic have been more than balanced by the advances resulting from their subsequent investigation.

CHAPTER TEN

Paradoxes in Higher Mathematics

THE mathematics involved in the previous chapters has for the most part been of the type ordinarily covered in secondary-school courses. Whenever such was not the case – as, for example, in the last three chapters – an attempt was made to develop as much of the mathematical background as was necessary for an understanding of the problem at hand.

This last chapter is designed for those whose work in mathematics has extended beyond the elementary level. It consists of some twenty paradoxes concerned chiefly with the subjects of trigonometry, analytical geometry, and calculus. A knowledge of these subjects will be assumed, and no attempt will be made to develop any of the necessary concepts or techniques. Furthermore, the solutions of the various problems will not be discussed in the main body of the text. The reader will thus be given a chance to diagnose the difficulties himself – a procedure he should follow if he wishes to derive the maximum amount of pleasure and profit from the chapter. Since, however, such a procedure might conceivably lead to insomnia, nervousness, and general irritability, a complete discussion of each problem is offered as usual in the Appendix.

GEOMETRY AND TRIGONOMETRY

Paradox 1. To prove that two non-parallel lines will never meet.[1]

Let *a* and *b* of Figure 102 be two non-parallel lines. Draw a third line *AB* so that it makes equal angles with *a* and *b*. It is evident at once that since angles 1 and 2 are obtuse, *a* and *b* can have no point in common to the left of the transversal *AB*. We need therefore consider only what happens to the right of *AB*.

Mark off $AC = BD = AB/2$. Points *C* and *D* cannot coincide as in Figure 103(a) – for if they did, the sum of two sides of the resulting triangle would be equal to the third side. Even less can the segments *AC* and *BD* have any other point in common – say the point *S* of Figure 103(b). For then the sum of two sides of the triangle *ABS* would be less than the third side.

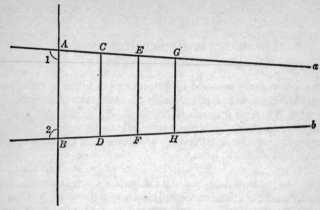

FIG. 102

Now draw CD and mark off $CE=DF=CD/2$. Reasoning exactly as before, we can show that the segments CE and DF have no point in common. We can therefore draw EF, mark off $EG=FH=EF/2$, and show that the segments EG and FH have no point in common. And so on. Since this same argument can be repeated indefinitely, we must conclude that the lines a and b will never meet.

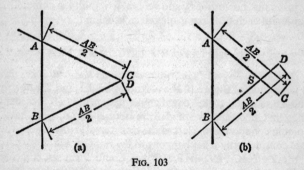

FIG. 103

Paradox 2. To prove that every triangle is isosceles.

In Figure 104 let ABC be any triangle, and let a, b, and c be the sides opposite the angles A, B, and C respectively. Extend BC a distance b to P, and AC a distance a to Q. Draw AP and BQ.

198

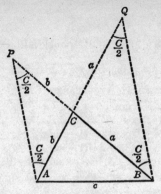

FIG. 104

In triangle APC, $AC=CP$, so $\angle CAP = \angle CPA$. Furthermore $\angle C$ of triangle ABC is an exterior angle of triangle APC. It follows that $\angle CAP = \angle CPA = \frac{1}{2}\angle C$. Similarly, $\angle CQB = \angle CBQ = \frac{1}{2}\angle C$. Now apply the law of sines to triangles ABP and ABQ. In the first of these,

$$\frac{BP}{AB} = \frac{a+b}{c} = \frac{\sin\left(A+\dfrac{C}{2}\right)}{\sin\dfrac{C}{2}}, \qquad (1)$$

and in the second,

$$\frac{AQ}{AB} = \frac{a+b}{c} = \frac{\sin\left(B+\dfrac{C}{2}\right)}{\sin\dfrac{C}{2}}. \qquad (2)$$

Therefore

$$\frac{\sin\left(A+\dfrac{C}{2}\right)}{\sin\dfrac{C}{2}} = \frac{\sin\left(B+\dfrac{C}{2}\right)}{\sin\dfrac{C}{2}}, \qquad (3)$$

or

$$\sin\left(A+\frac{C}{2}\right) = \sin\left(B+\frac{C}{2}\right), \qquad (4)$$

whence

$$A+\frac{C}{2} = B+\frac{C}{2}, \qquad (5)$$

199

and

$$A = B. \tag{6}$$

It follows that $a = b$, and that the triangle is isosceles by definition.

Paradox 3. To prove that $1 = 2$.[2]

We have, successively, for all values of x,

$$\cos^2 x = 1 - \sin^2 x, \tag{1}$$
$$(\cos^2 x)^{\frac{3}{2}} = (1 - \sin^2 x)^{\frac{3}{2}}, \tag{2}$$
$$\cos^3 x = (1 - \sin^2 x)^{\frac{3}{2}}, \tag{3}$$
$$\cos^3 x + 3 = (1 - \sin^2 x)^{\frac{3}{2}} + 3, \tag{4}$$
$$(\cos^3 x + 3)^2 = [(1 - \sin^2 x)^{\frac{3}{2}} + 3]^2. \tag{5}$$

Let x have the value $\pi/2$. Then $\cos x = 0$, $\sin x = 1$, and (5) reduces to

$$9 = 9,$$

a true result.

But now let x have the value π. Then $\cos x = -1$, $\sin x = 0$, and (5) reduces to

$$2^2 = 4^2.$$

That is to say, $2 = 4$, or $1 = 2$.

Paradox 4. To prove that $\sin x = 0$ for all values of x.[3]

As is well known, the power-series expression for $\sin x$ contains only odd powers of x. In other words, $\sin x$ can be written in the form

$$\sin x = a_1 x + a_2 x^3 + a_3 x^5 + a_4 x^7 + \dots \tag{1}$$

The coefficients $a_1, a_2, a_3, a_4, \dots$ can be determined in the following way. We know that

$$\sin^2 x = 1 - \cos^2 x,$$

whence

$$\sin x = (1 - \cos^2 x)^{\frac{1}{2}}. \tag{2}$$

Since $\cos^2 x \leqq 1$ for all values of x, the right-hand side of (2) can be expanded by the binomial theorem. This gives

$$\sin x = 1 - \tfrac{1}{2}\cos^2 x - \tfrac{1}{8}\cos^4 x - \tfrac{1}{16}\cos^6 x - \dots \tag{3}$$

But now think of the power-series expression for $\cos x$. It contains only even powers of x. Consequently the right-hand side of (3) contains only even powers of x. In other words, the coefficients of the odd powers of x in (3) are all zero. It follows that all the co-efficients a_1, a_2, a_3, a_4, ... (of 1) must vanish. That is to say, $\sin x = 0$ for all values of x.

Paradox 5. In solid geometry the traditional approach to the measurement of a right circular cylinder is through regular prisms. The lateral surface of such a cylinder, for example, is defined as the limit of the lateral surface of a regular inscribed prism as the number of lateral faces is indefinitely increased. Now it might be supposed that inscribed polyhedra other than regular prisms could be used equally well for the same purpose, provided, of course, that the number of faces be indefinitely increased and that the area of each face be made indefinitely small. Let us see whether or not this supposition is true in the following instance.

Consider a right circular cylinder whose radius is r and whose altitude is h. By means of planes parallel to the bases, divide the cylinder into $2n$ equal slices, each of altitude $h/2n$. One of these slices is shown in Figure 105. Divide the circumference of the lower base of this slice into $2m$ equal parts by the points A, B, C, D, E,

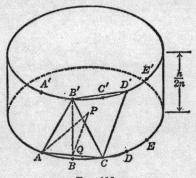

Fig. 105

... Divide the circumference of the upper base into $2m$ equal parts by the points A', B', C', D', E', ..., subject to the condition that A' lie on the same generator of the cylinder as A, B' on the same generator as B, C' on the same generator as C, and so on. Finally, construct the polyhedron whose faces are the isosceles triangles $AB'C$, $CB'D'$, $D'CE$, ... Do the same for each of the $2n$ slices.

Now denote the area of each triangular face by s, and the sum of the areas of all the triangular faces of the entire polyhedron by S. Since to each slice there corresponds $2m$ triangles, and since there are $2n$ slices, it follows that

$$S = 4mns. \qquad (1)$$

To express S in terms of m, n, r, and h, proceed as follows. Let P be the centre of the lower base, and let Q be the point of intersection of PB and AC. Then

$$s = AQ.B'Q. \qquad (2)$$

If we denote the angle APQ by α,

$$AQ = r \sin \alpha, \qquad (3)$$

and

$$
\begin{aligned}
B'Q &= \sqrt{(BB')^2 + (BQ)^2} \\
&= \sqrt{\left(\frac{h}{2n}\right)^2 + r^2(1 - \cos \alpha)^2} \\
&= \sqrt{\frac{h^2}{4n^2} + 4r^2 \sin^4 \frac{\alpha}{2}}.
\end{aligned} \qquad (4)
$$

Substituting (2), (3), and (4) in (1),

$$
\begin{aligned}
S &= 4mnr \sin \alpha \sqrt{\frac{h^2}{4n^2} + 4r^2 \sin^4 \frac{\alpha}{2}} \\
&= 2mr \sin \alpha \sqrt{h^2 + 16r^2n^2 \sin^4 \frac{\alpha}{2}}.
\end{aligned} \qquad (5)
$$

Finally, noting that $2m\alpha = 2\pi$, or that $m = \pi/\alpha$, (5) can be written in the form

$$S = 2\pi r \frac{\sin \alpha}{\alpha} \sqrt{h^2 + 16r^2n^2 \sin^4 \frac{\alpha}{2}}. \qquad (6)$$

We are now in a position to consider the limit of S as m and n tend to infinity.

(a) Let $n = km = k\pi/\alpha$, where k is any fixed constant. Then certainly m and n tend to infinity as α tends to zero. Moreover, the second quantity under the radical sign of equation (6) assumes a form which can be easily handled. That is to say,

$$16r^2n^2\sin^4\frac{\alpha}{2}=r^2k^2\pi^2\alpha^2\frac{\sin^4\frac{\alpha}{2}}{\left(\frac{\alpha}{2}\right)^4}. \tag{7}$$

Now the ratio of sin x to x tends to unity as x tends to zero. Hence the right-hand side of (7), because of the factor α^2, tends to zero as α tends to zero. Substituting (7) in (6) and passing to the limit, we have

$$\lim_{\alpha\to0} S=2\pi rh.$$

(b) But now suppose we let $n=km^2=k\pi^2/\alpha^2$. Then

$$16r^2n^2\sin^4\frac{\alpha}{2}=r^2k^2\pi^4\frac{\sin^4\frac{\alpha}{2}}{\left(\frac{\alpha}{2}\right)^4}. \tag{8}$$

Substituting (8) in (6) and passing to the limit,

$$\lim_{\alpha\to0} S=2\pi r\sqrt{h^2+r^2k^2\pi^4}.$$

(c) Finally, if we let $n=km^3=k\pi^3/\alpha^3$, we have

$$16r^2n^2\sin^4\frac{\alpha}{2}=r^2k^2m^2\pi^4\frac{\sin^4\frac{\alpha}{2}}{\left(\frac{\alpha}{2}\right)^4}, \tag{9}$$

and the quantity on the right, because of the factor m^2, tends to infinity as α tends to zero. If, then, we substitute (9) in (6) and pass to the limit,

$$\lim_{\alpha\to0} S=\infty.$$

The radically different results obtained in cases (a), (b), and (c) are surprising, for in all three cases m and n were made to tend to infinity together. It is true that in (a) we took n as a multiple of m, in (b) as a multiple of m^2, and in (c) as a multiple of m^3, but it seems incredible that these slight differences can lead to such tremendous differences in the results. There is, of course, no fallacy in the argument here – nothing is violated except intuition. But the example shows emphatically that we cannot, without careful attention to details, define a curved surface as the limit of the surface of an inscribed polyhedron with increasingly many faces.

ANALYTICAL GEOMETRY

Paradox 1. To prove that $\pi = \frac{8}{3}$.[5]

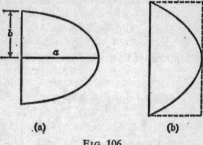

FIG. 106

The following two theorems are well known in the theory of conic sections.

I. The area of the semi-ellipse in diagram (a) of Figure 106 is $\pi ab/2$, where $2a$ and $2b$ are the major and minor axes, respectively, of the ellipse.

II. The area of the parabolic segment in diagram (b) of Figure 106 – a segment cut off by a chord perpendicular to the axis of the parabola – is $\frac{2}{3}$ that of the circumscribed rectangle.

If now the major axis of the ellipse is allowed to increase without limit, the ellipse degenerates to a parabola, and the semi-ellipse becomes a parabolic segment. But theorems I and II above are true regardless of the dimensions of the curves. Therefore

$$\frac{\pi ab}{2} = \frac{2}{3}(a \cdot 2b)$$

$$= \frac{4ab}{3}.$$

Hence

$$\frac{\pi}{2} = \frac{4}{3}, \text{ or } \pi = \frac{8}{3}.$$

Paradox 2. To prove that a diameter cuts a circle in only one point.[6]
The equations

$$x = \frac{1-t^2}{1+t^2}, \tag{1}$$

$$y = \frac{2t}{1+t^2} \tag{2}$$

are the parametric equations of a unit circle whose centre is at the origin. This statement is easily verified by noting that equations (1) and (2), if squared and added, reduce immediately to the equation $x^2 + y^2 = 1$.

Consider the intersection of the x-axis and the circle. That is to say, substitute $y = 0$ in (2). Equation (2) then reduces to $t = 0$, which, substituted in (1), gives $x = 1$. Therefore the x-axis cuts the circle only at the point $(1, 0)$.

Now by a proper choice of units and axes, any given circle can be made a unit circle, and any diameter of the given circle can be made to coincide with the x-axis. Hence any diameter of any circle cuts the circle in only one point.

Paradox 3. Consider the following problem.[7] A point P in three-dimensional Euclidean space is to be made collinear with two given points A and B. How many algebraic conditions must be imposed on the co-ordinates of P?

(*a*) Let Q and R be two arbitrary points subject only to the condition that A, B, Q, and R be not coplanar. Then if P is to be collinear with A and B, it is necessary and sufficient that P be coplanar with A, B, and Q, and also with A, B, and R. Therefore *two* conditions are imposed on the co-ordinates of P.

(*b*) Of the three distances AB, BP, and AP, it is necessary and sufficient that $BP + AP = AB$ or $AP + AB = BP$ or $AB + BP = AP$ – i.e. that

$$(-AB + BP + AP)(AB - BP + AP)(AB + BP - AP) = 0. \tag{1}$$

The left-hand side of this equation can be rationalized by multiplying both sides by the non-vanishing factor $-(AB + BP + AP)$. If the resulting four factors on the left are multiplied together, equation (1) assumes the form

$$(AB)^4 + (BP)^4 + (AP)^4 - 2(BP)^2(AP)^2$$
$$- 2(AP)^2(AB)^2 - 2(AB)^2(BP)^2 = 0. \tag{2}$$

The left-hand side of this equation is an unfactorizable rational expression in the co-ordinates of P. Hence only *one* condition is imposed.

Which of these two solutions is correct?

DIFFERENTIAL CALCULUS

Paradox 1. To prove that any two numbers are equal to each other.[8]

Let us start with the relation

$$x = a - b. \tag{1}$$

If we multiply both sides of (1) by x,

$$x^2 = ax - bx. \tag{2}$$

And if we square both sides of (1),

$$x^2 = a^2 - 2ab + b^2. \tag{3}$$

From (2) and (3),

$$ax - bx = a^2 - 2ab + b^2.$$

That is,

$$ax - a^2 + ab = bx - ab + b^2,$$

or

$$a(x - a + b) = b(x - a + b). \tag{4}$$

If we divide both sides of (4) by the factor $(x - a + b)$, we obtain $a = b$. But such an argument is obviously fallacious, for, since $x = a - b$, we are dividing both sides by zero. Very well, then, let us write

$$a \frac{(x - a + b)}{(x - a + b)} = b \frac{(x - a + b)}{(x - a + b)}. \tag{5}$$

Now when x has the value $a - b$, equation (5) reduces to $a(0/0) = b(0/0)$. In order to evaluate the indeterminate quantity $0/0$, we resort to a device frequently used for this purpose. That is to say, we make use of the fact that

$$\lim_{x \to \infty} \frac{f(x)}{g(x)} = \lim_{x \to \infty} \frac{f'(x)}{g'(x)}$$

If, then, we differentiate the numerators and denominators of the fractions in (5), we obtain

$$a \left(\frac{1}{1} \right) = b \left(\frac{1}{1} \right), \text{ or } a = b.$$

Paradox 2. To prove that all proper fractions have the same value.[9]

Let m and n be any two integers such that n is less than m. Then by ordinary long division,

$$\frac{1-x^n}{1-x^m}=1-x^n+x^m-x^{n+m}+x^{2m}-\ldots \tag{1}$$

Now let x have the value 1. The left-hand side of (1) assumes the indeterminate form 0/0. We can get round this difficulty by differentiating numerator and denominator before passing to the limit. We then have

$$\lim_{x\to 1}\frac{1-x^n}{1-x^m}=\lim_{x\to 1}\frac{-nx^{n-1}}{-mx^{m-1}}=\frac{n}{m}.$$

But the limit, as x approaches 1, on the right-hand side of (1) is $1-1+1-1+1-1+\ldots$ Therefore n/m, being equal to an expression which is independent of m and n, must always have the same value.

Paradox 3. Consider the triangle ABC of Figure 107. Suppose that AB is 12 inches long, and that the altitude CD is 3 inches long. Let us propose to find that point P on CD for which the sum of the distances of P from the three vertices is a minimum.[10]

If we denote by S the sum of the distances of P from A, B, and C, and by x the length of DP, then the problem is that of finding the value of x that makes S a minimum. Now,

$$S=CP+AP+PB.$$

But $CP=3-x$, and $AP=PB=\sqrt{x^2+36}$. Therefore

$$S=3-x+2\sqrt{x^2+36},$$

and

$$\frac{dS}{dx}=-1+\frac{2x}{\sqrt{x^2+36}}.$$

Making dS/dx zero gives $x=2\sqrt{3}=3\cdot464$, and for this value of x, P lies outside the triangle on DC produced. Hence there is no

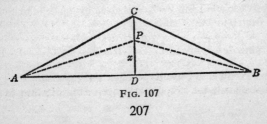

Fig. 107

point on *CD* for which *S* is a minimum. Yet the problem appears to be straightforward enough. What is wrong?

Paradox 4. To prove that every ellipse is a circle.[11]

Denote by *a* and *e*, respectively, the semi-major-axis and the eccentricity of the ellipse shown in Figure 108. It is well known that the length of the radius vector, drawn from the focus *F* to any point *P* of the ellipse, is given by the expression

$$r = a + ex.$$

Now $dr/dx = e$, and since there are no values of *x* for which dr/dx vanishes, *r* has no maximum or minimum. But the only closed

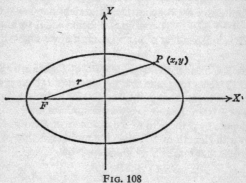

FIG. 108

curve in which the radius vector has no maximum or minimum is the circle. Therefore every ellipse is a circle.

INTEGRAL CALCULUS

Paradox 1. To prove that sin x=0 for all values of x.[12]

We know that $\sin 0 = 0$, and also that $\sin 2n\pi = 0$ for all integral values of *n*. Hence the area bounded by the curve $y = \sin x$ and the *x*-axis, between $x = 0$ and $x = 2n\pi$, is given by the definite integral of $\sin x$ from 0 to $2n\pi$. That is to say,

$$A = \int_0^{2n\pi} \sin x \, dx = \left[-\cos x \right]_0^{2n\pi} = -(\cos 2n\pi - \cos 0) = -1 + 1 = 0.$$

But if there is no area between $y = \sin x$ and the *x*-axis, the curve must coincide with the axis. Hence $\sin x = 0$ for all values of *x*.

Paradox 2. To prove that $-1=+1$.
We have

$$\int \frac{dx}{x} = \int \frac{-dx}{-x}. \tag{1}$$

Performing the indicated integration on both sides of (1),

$$\log x = \log (-x), \tag{2}$$

or

$$x = -x,$$

or

$$1 = -1.$$

Paradox 3. To prove that $\tan x = \pm i$ *for all values of* x.[13]
Consider the integral

$$I = \int \sin x \cos x \, dx.$$

If we think of $\cos x \, dx$ as $d (\sin x)$, then

$$I = \int \sin x \, d (\sin x) = \tfrac{1}{2} \sin^2 x. \tag{1}$$

If, on the other hand, we think of $\sin x \, dx$ as $-d (\cos x)$, then

$$I = -\int \cos x \, d (\cos x) = -\tfrac{1}{2} \cos^2 x. \tag{2}$$

From (1) and (2),

$$\sin^2 x = -\cos^2 x. \tag{3}$$

Dividing both sides of (3) by $\cos^2 x$, we get

$$\tan^2 x = -1.$$

Therefore

$$\tan x = \pm \sqrt{-1} = \pm i.$$

Paradox 4. To prove that an infinite area may generate a solid of revolution whose volume is finite.

The shaded area of Figure 109 is the area under the curve $y = 1/x$ from $x=1$ to $x=k$. This area, a function of k, is evaluated as follows:

$$A(k) = \int_1^k \frac{dx}{x} = \Big[\log x \Big]_1^k = \log k \text{ square units.}$$

I

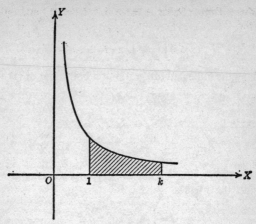

FIG. 109

If the area is revolved about the x-axis, it generates a solid whose volume is

$$V(k) = \pi \int_1^k \frac{dx}{x^2} = \left[-\frac{\pi}{x} \right]_1^k = \pi \left(1 - \frac{1}{k} \right) \text{ cubic units.}$$

Let k tend to infinity, Then

$$\lim_{k \to \infty} A(k) = \lim_{k \to \infty} (\log k) = \infty,$$

whereas

$$\lim_{k \to \infty} V(k) = \lim_{k \to \infty} \left[\pi \left(1 - \frac{1}{k} \right) \right] = \pi \text{ cubic units.}$$

COMPLEX NUMBERS

Paradox 1. To prove that $\pi = 0$.[14]

For all values of θ,

$$\cos \theta = \cos (2\pi + \theta),$$

and

$$\sin \theta = \sin (2\pi + \theta).$$

Therefore

$$\cos \theta + i \sin \theta = \cos (2\pi + \theta) + i \sin (2\pi + \theta),$$

and

$$(\cos \theta + i \sin \theta)^i = [\cos (2\pi + \theta) + i \sin (2\pi + \theta)]^i. \tag{1}$$

210

Recall from De Moivre's theorem that $(\cos x + i \sin x)^n = \cos nx + i \sin nx$. Hence (1) can be written in the form

$$\cos i\theta + i \sin i\theta = \cos i(2\pi + \theta) + i \sin i(2\pi + \theta). \qquad (2)$$

Now apply Euler's formula, $\cos x + i \sin x = e^{ix}$, to both sides of (2). We obtain

$$e^{-\theta} = e^{-2\pi - \theta}.$$

Dividing both sides of this expression by $e^{-2\pi - \theta}$,

$$e^{2\pi} = 1.$$

But e^x has the value 1 only when x is zero. Hence $2\pi = 0$, and $\pi = 0$.

Paradox 2. To prove that $-1 = +1$.[15]
Let x satisfy the equation $e^x = -1$. Square both sides. Then $e^{2x} = 1$. Now, as was noted only a few lines above, e^{2x} is 1 only when $2x$ is zero. Hence $2x = 0$, and $x = 0$. Substitute this value of x in the original equation. Then $e^0 = -1$. But any number raised to the 0th power is $+1$. In particular, $e^0 = +1$. Consequently $-1 = +1$.

Paradox 3. To prove that $-1 = +1$.
Consider the equation $(-1)^2 = +1$. Take the logarithm of both sides. Then $\log (-1)^2 = \log (1) = 0$. But $\log (-1)^2 = 2 \log (-1)$. Therefore $2 \log (-1) = 0$, and $\log (-1) = 0$. Consequently $\log (-1) = \log (1)$, or $-1 = +1$.

Paradox 4. Consider the following two linear homogeneous complex equations:

$$\begin{aligned}(a+bi)(p+qi) + (c+di)(r+si) &= 0, \\ (a'+b'i)(p+qi) + (c'+d'i)(r+si) &= 0.\end{aligned} \qquad (1)$$

How many conditions must be fulfilled if the equations (1) are to be compatible?[16]

(*a*) It is necessary and sufficient that the determinant of the coefficients vanish – that is, that

$$\begin{vmatrix} (a+bi) & (c+di) \\ (a'+b'i) & (c'+d'i) \end{vmatrix} = 0.$$

This complex equation is equivalent to the two real equations.

$$ac' - a'c = bd' - b'd$$
$$ad' + bc' = a'd + b'c. \tag{2}$$

(b) The equations (1) are equivalent to the system

$$ap - bq + cr - ds = 0$$
$$bp + aq + dr + cs = 0$$
$$a'p - b'q + c'r - d's = 0 \tag{3}$$
$$b'p + a'q + d'r + c's = 0.$$

But in order that the equations (3) be compatible it is necessary and sufficient that

$$\begin{vmatrix} a & -b & c & -d \\ b & a & d & c \\ a' & -b' & c' & -d' \\ b' & a' & d' & c' \end{vmatrix} = 0. \tag{4}$$

This determinant yields, of course, a single real equation.

Which of these two solutions is correct, the one that results in *two* equations, or the one that results in a *single* equation?

Appendix

CHAPTER 2

Pages 21–3

Paradox 1. It is incorrect to assume that 'the tail wind on the way south will speed up the plane to the same extent that the head wind will retard it on the way north'. Here we are again trying to get an average rate by averaging two rates maintained over equal *distances*. To analyse the problem, call the speed of the wind 50 miles per hour. Then the speed of the plane from London to Liverpool is 100+50, or 150 miles per hour; from Liverpool to London, 100−50, or 50 miles per hour. Hence the times for the trips down and back are $\frac{200}{150}$, or $\frac{4}{3}$, and $\frac{200}{50}$, or 4, hours respectively. It follows that the total time for the round trip is $5\frac{1}{3}$ hours, and that the average speed of the plane, in miles per hour, is

$$\frac{\text{total distance}}{\text{total time}} = \frac{400}{\frac{4}{3}+4} = \frac{400}{\frac{16}{3}} = \frac{400 \times 3}{16} = 75.$$

Paradox 2. The apple women made the error of calculating their average price rate by averaging their individual rates of 2 apples a penny and 3 apples a penny over the same number of *apples*. To guarantee the same receipts as those of the first day, they should have determined their price by dividing the total number of apples by the total number of pence – that is, $\frac{60}{25} = \frac{12}{5}$ apples a penny. They actually sold the apples at the rate of $2\frac{1}{2}$ apples a penny. There's where the missing penny went.

Paradox 3. The actual strokes occupy no appreciable length of time – the 5 seconds are accounted for by the 5 intervals between the 6 strokes. Between 12 strokes there are 11 intervals. Hence the correct answer is about 11 seconds.

Paradox 4. If the cost of the bottle were 1*s*, and that of the cork 1*d*, then the bottle would cost only 11*d* more than the cork. Second thought readily supplies the correct answer: 1*s* $\frac{1}{2}$*d*.

Paradox 5. At least, the answer is not 30 hours unless the frog is so stupid that he doesn't know when he is well out of a well. At the end of 27 hours he is 3 feet from the top. During the 28th hour he climbs the remaining 3 feet, and he's out.

Paradox 6. Need we point out that if the lengths of the trains are neglected, then slow train and express are the *same* distance from London when they meet?

CHAPTER 3

Pages 36–7

It was pointed out in Chapter 1 that in mathematics we are not concerned with the 'truth' of our definitions or assumptions, but only with their consistency. The fact that any number (other than 0) raised to the 0th power is defined as 1 is a case in point. It is easy to visualize a^2 as the product of two a's, a^3 as the product of three a's, and so on. But what is to be done with a^0? We obviously cannot visualize the product of zero – or no – a's. Now recall that if a is any number, and if m and n are positive whole numbers, then $a^m.a^n=a^{m+n}$. For example, $5^3.5^4=(5.5.5)(5.5.5.5)=5.5.5.5.5.5.5=5^7=5^{3+4}$. If we substitute 0 for m in this rule, we obtain $a^0.a^n=a^{0+n}=a^n$. But if $a^0.a^n=a^n$, we can divide both sides of this equation by a^n and obtain $a^0=1$. *Hence we define a^0 as 1 for the sake of consistency in our mathematical processes.* On the other hand, we cannot so define a^0 if $a=0$, for the last step would involve division by 0^n, or 0. This particular point is discussed at length in Chapter 5.

CHAPTER 5

Pages 86–7

Paradox 1. In step (1) it was assumed that $a=b+c$, or that $a-b-c=0$. We divided by $a-b-c$, or 0, to get equation (5).

Paradox 2. The left-hand side of each of the identities assumes the value $\frac{0}{0}$ when 1 is substituted for x. This problem serves as additional evidence that $\frac{0}{0}$ can be 'any number'.

Paradox 3. A case of division by zero in false whiskers. By adding 10 to the left-hand side, we changed the value of x to -7. Both sides of the equation were divided by $x+7$—now 0—in step (6).

Pages 89–90

Paradox 1. If the given equation is solved for x, it is found that $x=a+b$. Consequently the numerators in our first result, although equal, are zero – insufficient grounds for assuming that the denominators are equal. The second result is disposed of similarly.

Paradox 2. The solution for x is $x=a-b$. Hence the fraction $(3x-3a+3b)/(3x-3a+3b)$ is of the form $\frac{0}{0}$.

Paradox 3. Here the solutions for x and y are

$$x=a+b-c,$$
$$y=a-b+c.$$

These values reduce the fraction $(x-a-b+c)/(y-a+b-c)$ to the form $\frac{0}{0}$.

APPENDIX

Page 92

This paradox arises from contradictions in the original equations, which might have been written in factorized form as

$$(x-y)(2x-y)=4,$$
$$(x-y)(x+3y)=9.$$

It is now evident at once that these equations are not satisfied when x and y are equal.

Page 93

In passing from (5) to (6), only the positive signs were taken with the square roots. There is no contradiction if the negative sign is taken on the right, for then

$$n+1-\left(\frac{2n+1}{2}\right)=-n+\left(\frac{2n+1}{2}\right), \text{ or } \frac{1}{2}=\frac{1}{2}.$$

Pages 94–5

Paradox 1. It is assumed in step (1) that $a>b$. In step (4) both sides of the inequality are divided by $b-a$, a negative quantity.

Paradox 2. The logarithm of any number between 0 and 1 is negative, and in step (2) both sides of the inequality are multiplied by $\log\left(\frac{1}{2}\right)$.

Pages 96–7

Paradox 1. The error occurs in (3). Let's use our i's. Then (1) becomes $i=i$, and (3), $1/i=i/1$. Now (1) is true and (3) is false. For if (3) were true we should have, on clearing of fractions, $i^2=1$, whereas actually $i^2=-1$. In passing from (2) to (3) we attempted to apply to imaginary numbers the ordinary rule for division of radicals: $\sqrt{a/b}=\sqrt{a}/\sqrt{b}$.

Paradox 2. This is one of the more insidious of the fallacies. The trouble all occurs in step (1). It seems reasonable enough to argue that $\sqrt{x-y}=\sqrt{(-1)(y-x)}=i\sqrt{y-x}$, but this statement is valid only when $x-y$ is negative, so that $y-x$ is positive. It is perhaps simpler if put this way: we have said that any imaginary number such as $\sqrt{-a}$ (and if this number is imaginary, a must be positive) can be written as $i\sqrt{a}$. We have *not* said that the real number $\sqrt{a}$ can be written as $i\sqrt{-a}$, for this would give $i^2 \cdot \sqrt{a}$, or $-\sqrt{a}$, an immediate contradiction. In the problem under consideration we can assume that a and b are not equal, for if they were equal, step (5) would involve division by zero. Then either $a>b$ or $b>a$, which means that one or the other of the left-hand sides of (2) and (3) is a real number, and this fact invalidates the whole argument.

215

CHAPTER 6
Pages 102–6

Paradox 1. If the diameters are properly drawn, then the line *PS* will not cut the circles in two distinct points *M* and *N*, but will pass through *R*, as shown in Figure 110. To prove this, draw the diameters and con-

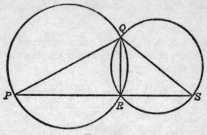

Fig. 110

nect *R* with *P*, *S*, and *Q*. Since angles *PRQ* and *SRQ* are inscribed in semi-circles, they are right angles, But their sum must therefore be a straight angle, and this makes *PRS* a straight line. Finally, since between the two points *P* and *S* only one straight line can be drawn, this line must go through *R*.

Paradox 2. In Figure 111, properly drawn, it is readily seen that the line *PE* falls *outside* the rectangle. Our proof for the equality of angles *DAP* and *EBP* is still valid, but it is now evident that the right angle *CBG* is no longer a part of the angle *EBP*.

Paradox 3. Similar to Paradox 2. If the figure were drawn correctly, it would be found that *EK* lies completely outside triangle *ABC*. Although our proof that $\angle DBK$ and $\angle EBK$ are equal is still valid, the 60° angle *ABC* is no longer a part of $\angle EBK$.

Paradox 4. Similar to Paradox 2. The perpendicular bisectors actually meet outside the quadrilateral, as in Figure 66(b), but in such a way that the line *OB* lies completely outside as well. Hence, although $\angle AOP = \angle 1 - \angle 3$, $\angle BOP = \angle 2 + \angle 4$, not $\angle 2 - \angle 4$.

Paradox 5. Consider the original proportion $AB/BC = AD/DC$. Since *B* is the internal point of division of *AC*, and *D* the external point, it is evident at once that *AD* must be greater than *AB*. It follows (see page 94) that *DC* must be greater than *BC*. But in that case *Q*, the mid-point of *BD*, must lie outside the circle, so that the perpendicular bisector of *BD* does not intersect the circle at all. In other words there is no point *P*. Our proof breaks down completely when we first use *P* in step (8).

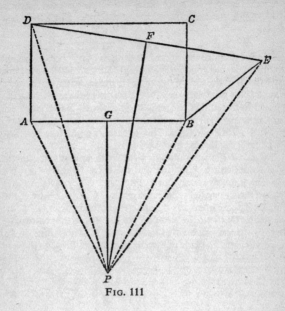

FIG. 111

Pages 109–11

Paradox 1. The fallacy here is a disguised case of division by zero. We conclude from step (8) that the denominators must be equal because the numerators are equal. But the numerators are zero, and this fact invalidates the conclusion. (See page 88.) To show that the numerators are zero, note that the triangles ABC and ADC are similar. Hence $AC/AD = AB/AC$, or $\overline{AC^2} = AB \cdot AD$. That is, $\overline{AC^2} - AB \cdot AD = 0$.

Paradox 2. Suppose we solve (1) and (2) for r and t in terms of p, q, and s. From the equations in question we have

$$pr - qt = -ps, \tag{a}$$
$$qr - pt = ps. \tag{b}$$

Adding (a) and (b), we have

$$(p+q)r - (p+q)t = 0,$$

or

$$(p+q)(r-t) = 0.$$

The last equation will be satisfied if *either* of the two factors is zero. In (4) we disregarded the possibility that $r-t$ might be zero, and so had to

217

conclude that $p+q$ is zero. Since $p+q$ is *not* zero, $r-t$ *must be* zero. Then, in (4), the fraction $(t-r)/(r-t)$ becomes $\frac{0}{0}$, which is meaningless.

Pages 114–15

Paradox 1. Since the sum of the angles of a spherical triangle can be anything between (but not including) 180° and 540°, we cannot assume, as we did, that the sum of the angles is the *same* (that is to say, x) for any triangle. Nor is it true to say that 'the sum of the angles of triangle ABC is equal to the sum of the angles of the three small triangles minus the sum of the angles at P,' for this implies the assumption that (Figure 72) $\angle CAP + \angle PAB = \angle CAB$. This is certainly not true when applied to the angles of spherical triangles.

Paradox 2. This paradox is a rather ingenious one, for the correct figure is difficult to visualize, and the correct analysis is a bit lengthy, although it involves only fairly simple ideas. Let us fix our attention on PA and the associated sphere and circle. To fix the centre of the circle in which the plane intersects the sphere whose diameter is PA, drop a perpendicular OQ from the mid-point of PA—from the centre of the sphere, that is – to the plane m, as in Figure 112. (The radius of a sphere, if perpendicular to a plane of intersection, passes through the centre of the

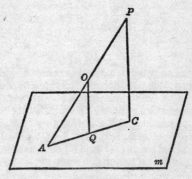

Fig. 112

circle of intersection.) Then with Q as centre and QA as radius draw the circle in m. But now drop a perpendicular PC from P to m and draw QC. Since AQC is the projection of the line PA on the plane, AQC will be a straight line. Furthermore, since AO and OP are radii of the same sphere, they are equal. But OQ is parallel to PC. (Two lines perpendicular to the same plane are parallel.) Therefore $AQ=QC$. (A line parallel to one side of a triangle divides the other two sides proportionally.) But this result means that C must lie on the circle of intersection, and AQC must be a diameter.

If now we treat in the same way the circle formed by the plane and the sphere about PB as diameter, our situation in plane m will appear as in Figure 113. Draw AD and DB. We have already proved that if AC and

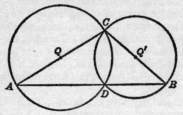

FIG. 113

BC are diameters, ABD must be a straight line. See discussion of Paradox 1, Ch. 6, page 216.) Now return to our original proof. We argued that since PC is perpendicular to the two intersecting straight lines AC and CB, it is perpendicular to plane m. True enough. But we applied the same argument to PD, thinking of AD and DB as *intersecting* straight lines. Since AD and DB are parts of one and the same line, PD need not be perpendicular to m, even though it is perpendicular to the line ADB. Finally, for different choices of A and B all the circles of intersection pass through C, so that PC is a perpendicular common to all choices. The second intersection D is a varying point. In each case PD is perpendicular to the corresponding line ADB, but not to plane m.

CHAPTER 7

Pages 134–6

Paradox 1. The length of the limiting line *appears* to be $\sqrt{2}$ only because the limiting line *appears* to be the hypotenuse of the right triangle. Consider the first line, L_1, shown in diagram (b) of Figure 79. It is evident at once that the sum of the horizontal segments is 1, and that the sum of the vertical segments is also 1. Hence the length of L_1 is 2. But the same argument applies to L_2 and L_3, shown in diagrams (c) and (d). That is to say, in each of these cases the sum of the horizontal segments is 1 as is the sum of the vertical segments. Hence L_2 and L_3, like L_1, are each 2 units long, Now regardless of how many times the number of 'steps' is doubled and redoubled, the sum of the horizontal segments remains 1, and the sum of the vertical segments remains 1. Consequently *every one* of the broken lines $L_1, L_2, L_3, L_4, L_5, L_6, \ldots$ is 2 units long. It follows that the length of the limiting line is also 2 units, and not $\sqrt{2}$ units.

Paradox 2. Consider the curve C_1, shown in Figure 80(a). Since the circumference of a circle is equal to π times the diameter of the circle, the length of C_1 is $\pi(AB)$. The curve C_2, in diagram (b), consists of two circles, each of diameter $(AB)/2$. The circumference of each circle is $\pi(AB)/2$. Therefore the length of C_2 is $2 . \pi(AB)/2$, or $\pi(AB)$. The curve C_3, in diagram (c), consists of four circles, each of diameter $(AB)/4$. The circumference of each circle is $\pi(AB)/4$. Therefore the length of C_3 is $4 . \pi(AB)/4$, or $\pi(AB)$. Similarly, the length of C_4 is $8 . \pi(AB)/8$, or $\pi(AB)$, And so on. Consequently the length of *every one* of the curves C_1, C_2, $C_3, C_4, C_5, C_6, \ldots$ is $\pi(AB)$. It follows that the length of the limiting curve is not $2(AB)$, but $\pi(AB)$.

Paradox 3. Consider the curve C_1, shown in Figure 81(a). This curve consists of four semicircles, each constructed on a side of the inscribed square. Since the circumference of a circle is equal to π times the diameter of the circle, the length of a semicircle is equal to $\frac{1}{2}\pi$ times the diameter of the semicircle. Hence the length of each semicircle in the figure is equal to $\frac{1}{2}\pi$ times one side of the square, and the length of the four semicircles is equal to $\frac{1}{2}\pi$ times the sum of the four sides of the square. If we denote the perimeter of the square by p_1, we can express the length of C_1 compactly as $\pi . p_1/2$. Again, C_2 is made up of eight semicircles, each constructed on one of the eight sides of the inscribed octagon of diagram (b). Hence the length of C_2 is $\frac{1}{2}\pi$ times the sum of the eight sides of the octagon. Or, if we denote the perimeter of the octagon by p_2, we can express the length of C_2 as $\pi . p_2/2$. In the same way the lengths of C_3 and C_4 can be expressed as $\pi . p_3/2$ and $\pi . p_4/2$, where p_3 and p_4 denote, respectively, the perimeters of the sixteen-sided and thirty-two-sided inscribed polygons of diagrams (c) and (d). Therefore the lengths of the successive curves C_1, C_2, C_3, C_4, C_5, C_6, ... are respectively,

$$\frac{\pi}{2} . p_1, \frac{\pi}{2} . p_2, \frac{\pi}{2} . p_3, \frac{\pi}{2} . p_4, \frac{\pi}{2} . p_5, \frac{\pi}{2} . p_6, \ldots,$$

where $p_1, p_2, p_3, p_4, p_5, p_6, \ldots$ denote the perimeters of the successive inscribed polygons. But in introducing the notion of a limiting curve in general, it was pointed out that it can be proved rigorously that the sequence of inscribed polygons approaches the circle as a limit. That is to say, the limit of the sequence of perimeters $p_1, p_2, p_3, p_4, p_5, p_6, \ldots$ is the circumference of the circle, or $2\pi R$. Consequently the length of the limiting curve (which *appears* to approach the circumference of the circle) is not $2\pi R$, but $\frac{1}{2}\pi$ times $2\pi R$, or $\pi^2 R$.

Page 154

Ever since Cantor's discovery of the transfinite numbers A_1 and C, mathematicians have been trying to find an infinite class whose transfinite number is greater than A_1, and less than C. All such attempts have been in vain. The question then arises whether or not the *assumption* that there is no transfinite number between A_1 and C is a consistent one – that is to say, whether or not it will ever lead to contradictory results. K. Gödel, an Austrian logician, has succeeded in proving the following theorem:

If the ordinary axioms – or assumptions – of the theory of aggregates are consistent, then the ordinary axioms, together with the assumption that there is no transfinite number between A_1 and C, are also consistent.

It is interesting to note further, that Gödel *conjectures* that the denial of the assumption in question would also be consistent. If this conjecture can ever be *proved*, it will mean that it will never be possible to decide, by means of the ordinary methods of the theory of aggregates, whether or not there exists a transfinite number greater than A_1 and less than C.

CHAPTER 8

Page 165

Paradox 1. The first solution is wrong, the second right. The quickest way to settle the matter is to examine a diagram showing all possibilities. Such a diagram is given in Figure 114, from which it is easy to see that the three coins can be tossed in any one of 8 equally likely ways. Of these 8 ways only 2 – the first and eighth – are favourable. Therefore the correct probability is $\frac{2}{8}$, or $\frac{1}{4}$.

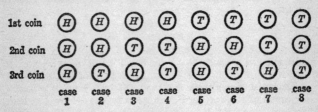

Fig. 114

In the incorrect solution we argued that two of the coins must come down alike. Let us suppose, to fix our ideas, that these are heads. We then assumed that it is just as likely for the third coin to be like the first two as to be unlike them. A glance at the figure will show that this assumption is not valid. Two (or more) heads appear in 4 of the 8 possible cases – the first, second, third, and fifth. In only *one* of these 4 cases are all three coins heads. Consequently it is *three times as likely* for the third coin to be unlike the other two as to be like them.

Paradox 2. Figure 115 shows that there are 6 possible results in the game. The marbles are shown diagrammatically – Peter's first marble on the left, his second on the right. The numbers on the marbles indicate

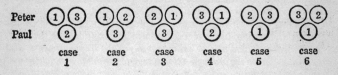

Peter	① ③	① ②	② ①	③ ①	② ③	③ ②
Paul	②	③	③	②	①	①
	case 1	case 2	case 3	case 4	case 5	case 6

FIG. 115

whether that particular marble got first, second, or third place in the game. In the fourth case, for example, Peter's second marble came first, Paul's marble second, and Peter's first marble third. Peter wins in 4 of the 6 cases – all but the last 2. Therefore the correct probability is $\frac{2}{3}$, and not $\frac{3}{4}$.

Let us see what is wrong with the second solution suggested. Here it was argued that the following 4 cases are the only ones possible. (i) Both of Peter's better than Paul's. (ii) Peter's first better than Paul's and his second worse. (iii) Peter's second better than Paul's and his first worse. (iv) Both of Peter's worse than Paul's. Now compare these 4 cases, labelled with Roman numerals, with the 6 cases of Figure 115, labelled with Arabic numerals. We see that (i) includes (2) and (3), that (ii) is the same as (1), that (iii) is the same as (4), and that (iv) includes (5) and (6). Since cases (1) to (6) are equally likely, cases (i) to (iv) are not. Case (iv) – the only one which makes Peter lose – is *more* likely than either case (ii) or case (iii).

CHAPTER 10
Pages 197–201

Paradox 1. The statement that a and b will 'never' meet is incorrect. Let us assume, in the figure which accompanies the problem, that $AB=1$. Denote each of the equal angles ABD and BAC by θ. Since $AC=BD=\frac{1}{2}$, the projection of either AC or BD on AB is $(\frac{1}{2})\cos\theta$. Now CD, being parallel to AB, is equal in length to its own projection on AB. It is therefore easy to see that $CD=1-\cos\theta$. Similarly, $EF=(1-\cos\theta)^2$, $GH=(1-\cos\theta)^3$, .. In general, the length of the nth line drawn between a and b is $(1-\cos\theta)^n$. As the construction is 'continued indefinitely', n tends to infinity. And, since $0<\cos\theta<1$,
$$\lim_{n\to\infty}(1-\cos\theta)^n=0.$$
Consequently a and b will 'ultimately' meet.

Paradox 2. In steps (4) and (5) we concluded, since

$$\sin \left(A+\frac{C}{2}\right)=\sin \left(B+\frac{C}{2}\right),$$

that

$$A+\frac{C}{2}=B=\frac{C}{2}.$$

This conclusion is not necessarily true. That is to say, if $\sin x = \sin y$, x is not necessarily equal to y, but may be equal to the supplement of y. This conclusion follows from the fact that

$$\sin y = \sin (180° - y).$$

Thus, in place of step (5) we may have

$$A+\frac{C}{2}=180°-\left(B+\frac{C}{2}\right).$$

Adding $B+(C/2)$ to both sides of this equation, we obtain
$$A+B+C=180°,$$
a true result.

Paradox 3. A case of failure to examine the double sign when extracting the square root in step (2). The left-hand side of (3) should be $\pm \cos^3 x$ in which cases (5) will read

$$(\pm\cos^3 x+3)^2=[(1-\sin^2 x)^{\frac{3}{2}}+3]^2. \tag{5'}$$

The negative value of the term $\pm\cos^3 x$ must be taken when x has the value π. The relation (5') then reduces to $4^2=4^2$, or $16=16$.

Paradox 4. The fallacy here is the same as that noted in Paradox 3. Steps (2) and (3) should read, respectively,

$$\sin x = \pm (1-\cos^2 x)^{\frac{1}{2}},$$
$$\sin x = \pm (1 - \tfrac{1}{2}\cos^2 x - \tfrac{1}{8}\cos^4 x - \tfrac{1}{16}\cos^6 x - ...).$$

With the double sign, $\sin x$ is no longer an 'even' function.

Pages 204–5

Paradox 1. A case of misuse of infinity. If the major axis of the ellipse is allowed to increase without limit, the area in question – call it either a semi-ellipse or a parabolic segment – becomes infinite, and the relation

$$\frac{\pi a b}{2}=\frac{2}{3}\,(a.2b),$$

is meaningless.

Paradox 2. The value of y is zero not only when t is zero, but also when t is infinite. To be more specific, equations (1) and (2) give

$$\lim_{t \to \infty} x = -1, \quad \lim_{t \to \infty} y = 0.$$

These observations account for the second point at which the x-axis intersects the circle – that is, the point $(-1, 0)$.

Paradox 3. The first solution is correct. In order to see how the single equation (2) reduces to two equations, proceed as follows. Choose a system of rectangular co-ordinates so that the origin is at the given point A, and so that the x-axis passes through B. The co-ordinates of A, B, and P can then be taken, respectively, as $(0, 0, 0)$, $(b, 0, 0)$, and (x, y, z). Equation (2) reduces at once to

$$y^2 + z^2 = 0,$$

or

$$y = 0 \quad \text{and} \quad z = 0.$$

Hence two conditions are imposed on the co-ordinates of P.

Pages 206–8

Paradox 1. The use of the theorem to the effect that

$$\lim_{x \to \alpha} \frac{f(x)}{g(x)} = \lim_{x \to \alpha} \frac{f'(x)}{g'(x)}$$

is not legitimate in this problem. The quantity x is not a variable, but a constant. At the very beginning of the problem it was assumed that $x = a - b$.

Paradox 2. The trouble here lies not in the application of the theorem used in Paradox 1, but in the expression to which the right-hand side of (1) reduces when x has the value 1 – that is to say, in the series $1 - 1 + 1 - 1 + 1 - 1 + \dots$ It was argued that this series always has the same value, the word 'value' presumably referring to the sum of the series. But the series is an oscillating series, and so has no definite sum. (Compare this paradox with that on page 124, in which the series in question is used to 'prove' that $\frac{1}{2} = \frac{1}{3} = \frac{1}{4} = \frac{1}{5} = \dots$)

Paradox 3. If P is to lie on CD, x can assume values only between 0 and 3. The function dS/dx vanishes for no value of x in this range. Therefore the value of x for which S is a minimum cannot be found by making dS/dx zero. For x in the range 0 to 3, S assumes a minimum value when x is 3. This fact can be verified by inspection either of the function S itself, or of its graph for $0 \leq x \leq 3$. It is true that the function S is a minimum for $x = 2\sqrt{3}$. But in this case the distance CP, which is denoted by $3 - x$, is negative.

Paradox 4. Similar to Paradox 3. The relation between r and x is linear, so there is obviously no value of x for which dr/dx vanishes. The permissible range of values for x runs from $-a$ to $+a$. It can readily be seen by inspection that r is a maximum when $x = +a$, and a minimum when $x = -a$.

APPENDIX

Pages 208–9

Paradox 1. For all integral values of n, it is true that $\sin 2n\pi = 0$, but it is also true that $\sin nx = 0$. The area bounded by $y = \sin x$ and the x-axis between 0 and π is equal numerically to the area bounded by the curve and the axis between π and 2π, but these two areas are opposite in sign. It is easy to see that the area obtained by integration from 0 to $2n\pi$ consists of an equal number of positive and negative portions, and that the algebraic sum of these portions is zero.

Paradox 2. The fallacy here lies in the fact that the constant of integration was overlooked. If two functions are equal, it does not follow that their integrals are equal – they may differ by a constant. Step (2) should read

$$\log x = \log(-x) + C.$$

The right-hand side of this relation reduces to the left-hand side if the value of C is taken as $\log(-1)$. That is to say,

$$\log(-x) + \log(-1) = \log(-x)(-1)$$
$$= \log x.$$

Paradox 3. Similar to Paradox 2. Substituting $1 - \cos^2 x$ for $\sin^2 x$ in (1),

$$I = \frac{1}{2}(1 - \cos^2 x)$$
$$= \frac{1}{2} - \frac{1}{2}\cos^2 x.$$

This result differs from the value of I as given in step (2) only by the constant $\frac{1}{2}$.

Paradox 4. There is no fallacy in the argument. The *area* under the curve is generated by the ordinate, $1/x$. As x increases without limit, $1/x$ tends to zero, but so slowly that the entire area is infinite. The *volume*, on the other hand, is generated by the cross section of the solid – a quantity proportional to $1/x^2$. As x increases without limit, the quantity $1/x^2$ tends to zero much more rapidly than $1/x$ – rapidly enough, as a matter of fact, to make the entire volume finite. (Compare with the fact that the series

$$\frac{1}{2} + \frac{1}{3} + \frac{1}{4} + \frac{1}{5} + \frac{1}{6} + \frac{1}{7} + \dots$$

diverges to infinity, whereas the series

$$\frac{1}{2^2} + \frac{1}{3^2} + \frac{1}{4^2} + \frac{1}{5^2} + \frac{1}{6^2} + \frac{1}{7^2} + \dots$$

converges to a finite limit.)

225

Pages 210–11

Paradox 1. De Moivre's theorem,

$$(\cos x + i \sin x)^n = \cos nx + i \sin nx,$$

is valid only for real values of n. In the paradox under consideration it was incorrectly assumed that this theorem could be applied when n has the value i. In addition, it is incorrect to argue that e^x has the value 1 only when x is zero. True enough for real values of x, but not for complex values. In order to verify this statement, substitute in Euler's formula,

$$e^{ix} = \cos x + i \sin x,$$

the value $x = 2n\pi$. It is seen at once that $e^{2n\pi i}$ has the value 1 for all integral values of n.

Paradoxes 2 and 3. Similar to Paradox 1. Paradox 2 is the exponential form, and Paradox 3 the logarithmic form, of one and the same argument. By Euler's formula, the equations $e^x = -1$ and $e^{2x} = 1$ of Paradox 2 are satisfied if $x = n\pi i$, where n is any integer. The equation $e^{2x} = 1$ does not necessarily imply that $2x$, and hence x, has the value zero.

Paradox 4. To show that the second solution reduces to the first, interchange the second and third rows of the determinant of equation (4). Then

$$\begin{vmatrix} a & -b & c & -d \\ a' & -b' & c' & -d' \\ b & a & d & c \\ b' & a' & d' & c' \end{vmatrix} = 0.$$

Now apply a theorem of Laplace on the development of determinants. The result is the equation

$$(ac' - a'c)^2 + (bd' - b'd)^2 + (ad' - a'd)^2 + (bc' - b'c)^2 - 2(a'b - ab')(c'd - cd') = 0.$$

This equation can be written in the form

$$[(ac' - a'c) - (bd' - b'd)]^2 + [(ad' + bc') - (a'd + b'c)]^2 = 0.$$

Since a, b, c, d, a', b', c', and d' are all real quantities, this *single* equation is equivalent to the two equations

$$ac' - a'c = bd' - b'd,$$
$$ad' + bc' = a'd + b'c.$$

These equations are, of course, the equations (2) reached in the first solution.

Notes and References

CHAPTER 2

1. Good source books for material of this sort are W. W. R. Ball, *Mathematical Recreations and Essays*, London (Macmillan), 1931 (10th ed.), and W. Lietzmann, *Lustiges und Merkwürdiges von Zahlen und Formen*, Breslau (Hirt), 1930 (4th ed.).

2. Lewis Carroll (C. L. Dodgson), *Further Nonsense*, New York (Appleton), 1926, pp. 91, 92.

3. Some of these examples are to be found in H. E. Dudeney, *Amusements in Mathematics*, London (Nelson), 1917, pp. 8, 9.

4. The author's attention has been called to the following actual instance of an even greater complication in the family of the second wife of Percy Bysshe Shelley, the famous English poet.

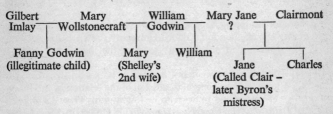

5. Deceased Wife's Sister Act of 1907, and Deceased Brother's Widow Act of 1921.

CHAPTER 3

1. W. W. R. Ball, *Mathematical Recreations and Essays*, London (Macmillan), 1931 (10th ed.), p. 229. This is Ball's version of de Parville's account in *La Nature*, Paris, 1884, part I, pp. 285, 286.

2. The author is indebted to H. Steinhaus for this neat way of presenting the largest prime. See his *Mathematical Snapshots*, New York (Stechert), 1938, p. 12. (See Additional Notes on p. 233.)

3. Information on such topics as Fermat's numbers, perfect numbers, and the division of the circle can be found in almost any history of mathematics. See, for example, D. E. Smith, *A History of Mathematics*, New York (Ginn), 1925. A good discussion of the first two of these three topics is to be found also in Ball, *op. cit.*, pp. 37–40.

4. F. Cajori, *A History of Elementary Mathematics*, New York (Macmillan), 1914, pp. 1–18.

5. For a complete discussion of the theory of this game – commonly called nim – see C. L. Bouton, *Annals of Mathematics*, series 2, vol. 3 (1901–2), pp. 35–9.

6. See, for example, Ball, *op. cit.*, pp. 4–13; also W. Lietzmann, *Lustiges and Merkwürdiges von Zahlen und Formen*, Breslau (Hirt), 1930 (4th ed.), pp. 153–69. Perhaps the best popular collection of mind-reading tricks involving numbers is to be found in R. V. Heath, *Mathemagic*, New York (Simon & Schuster), 1923.

CHAPTER 4

1. Compare W. W. R. Ball, *Mathematical Recreations and Essays*, London (Macmillan), 1931 (10th ed.), pp. 52–4. According to Ball, earliest reference to this paradox is *Zeitschrift für Mathematik und Physik*, vol. 13 (1868), p. 162. See also *American Mathematical Monthly*. R. C. Archibald, vol. 25 (1918), p. 236; and W. Weaver, vol. 45 (1938), p. 234.

2. See, for example, A. H. Church, *On the Interpretation of Phenomena of Phyllotaxis*, London (Oxford University Press), 1920.

3. The equation in polar co-ordinates of the logarithmic spiral is $r = a^\theta$, or $\theta = \log_a r$.

4. Jay Hambidge has written a number of books on dynamic symmetry. Perhaps the best general discussion of the relation of the Fibonacci series to nature and to art is to be found in his *Practical Applications of Dynamic Symmetry*, New Haven (Yale University Press), 1932. This book contains numerous illustrations of plant growths, shell spirals, and the like.

5. For a complete discussion of curves of constant breadth, see H. Rademacher and O. Toeplitz, *Von Zahlen und Figuren*, Berlin (Springer), 1930, pp. 128–41.

6. Galileo Galilei, *Dialogues Concerning Two New Sciences*, New York (Macmillan), 1914, pp. 20–6. This book is an English translation of the original Italian text, published in Leyden in 1638.

7. See, for example, W. W. R. Ball, *op. cit.*, pp. 170–81, for this problem and some of its generalizations.

8. This surface is discussed at length in D. Hilbert and S. Cohn-Vossen, *Anschauliche Geometrie*, Berlin (Springer), 1932, pp. 271–6.

9. Good photographs of the Möbius strip and other strips discussed here are to be found in H. Steinhaus, *Mathematical Snapshots*, New York (Stechert), 1938, pp. 112–16.

10. W. W. R. Ball, *op. cit.*, pp. 321–36. See also, by the same author, *String Figures*, Cambridge (Heffer), 1921 (2nd ed.).

11. Figure is from H. Steinhaus, *op. cit.*, p. 118.

CHAPTER 5

1. W. F. White, *A Scrap-Book of Elementary Mathematics*, Chicago (Open Court), 1910 (2nd ed.), p. 88.

2. J. R. D'Alembert, *Opuscules mathématiques*, Paris, 1761, vol. 1, p. 201.

3. W. Lietzmann, *Trugschlüsse*, Leipzig (Teubner), 1923 (3rd ed.), p. 8.

4. W. Lietzmann, *op. cit.*, p. 40.

5. W. F. White, *op. cit.*, p. 78.

6. W. Lietzmann, *op. cit.*, p. 14.

7. E. Gelin, *Mathesis*, vol. 13 (1893), p. 224.

8. W. Lietzmann, *op. cit.*, pp. 14, 15.

9. W. F. White, *op. cit.*, p. 84.

10. W. Lietzmann, *op. cit.*, pp. 9, 10.

11. The three following examples are from W. Lietzmann, *op. cit.*, pp. 12, 13.

12. See, for example, B. Russell, *Introduction to Mathematical Philosophy*, London (Allen & Unwin), 1919, pp. 1–19.

13. W. F. White, *op. cit.*, p. 85.

14. W. W. R. Ball, *Mathematical Recreations and Essays*, London (Macmillan), 1931 (10th ed.), p. 30. Attributed to G. T. Walker.

CHAPTER 6

1. See, for example, T. L. Heath, *The Thirteen Books of the Elements of Euclid*, Cambridge (Univ. Press), 1926 (2nd ed.), vol. 1, p. 7.

2. W. W. R. Ball, *Mathematical Recreations and Essays*, London (Macmillan), 1931 (10th ed.), p. 48. Ball's discussion is by no means as detailed as the one given here.

3. See, for example, Hawkes, Luby, and Touton, *New Plane Geometry*, New York (Ginn), 1917, p. 405.

4. W. W. R. Ball, *op. cit.*, p. 45.

5. W. W. R. Ball, *op. cit.*, p. 49.

6. M. Laisant, *Mathesis*, vol. 13 (1893), p. 224.

7. P. Stäckel, *Archiv der Mathematik und Physik*, series 3, vol. 12 (1907), p. 370.

8. *Preussische Lehrerzeitung*, about 1913.

9. M. Coccoz, *L'Illustration*, Paris, 12 Jan. 1895.

10. W. Lietzmann, *Trugschlüsse*, Leipzig (Teubner), 1923 (3rd ed.), pp. 32, 33.

11. W. Lietzmann, *op. cit.*, pp. 31, 32.

12. W. Lietzmann, *op. cit.*, pp. 35, 36.

13. G. Gille, *Mathesis*, vol. 29 (1909), p. 97.

CHAPTER 7

1. An exhaustive bibliography of researches concerning Zeno's paradoxes is to be found in an article by F. Cajori in *American Mathematical Monthly*, vol. 22 (1915), pp. 1–6, 292–7.

2. For a technical discussion of the convergence and divergence of infinite series, refer to any good text on the subject – for example, T. J. Bromwich, *An Introduction to the Theory of Infinite Series*, London (Macmillan), 1908.

3. The number e, an irrational number, is as important to calculus as the number π is to geometry. Its value to five decimal places is 2·71828. 'Log$e$ 2' signifies the logarithm of 2 to the base e – the power to which e must be raised if the resulting number is to be equal to 2. The proof of the convergence of the series in question is given in T. J. Bromwich, *op. cit.*, p. 51.

4. Bernard Bolzano, *Die Paradoxien des Unendlichen*, published posthumously, edited by Fr. Přihonsky, Leipzig (Reclam), 1851. Reprinted Leipzig (Meiner), 1920.

5. *Annales de mathématique*, vol. 20 (1830), p. 364. Article signed 'M. R. S.'

6. We shall see presently that the same series can be summed in other ways. W. W. R. Ball believes that this particular form of the paradox first appeared in his *Algebra*, Cambridge, 1890, p. 430.

7. For the proof of this theorem see, for example, T. J. Bromwich, *op. cit.*, pp. 68–70. Although Riemann proved the theorem in 1854, it was not published until 1867.

8. This form of the paradox is attributed to Dirichlet.

9. G. Chrystal, *Algebra*, Edinburgh, 1889, vol. 2, p. 159.

10. W. Lietzmann, *Trugschlüsse*, Leipzig (Teubner), 1923 (3rd ed.), p. 43.

11. Galileo Galilei, *Dialogues Concerning Two New Sciences*, New York (Macmillan), 1914, pp. 27–9. This book is a translation of the original Italian text, published in Leyden in 1638.

12. *Journal für Mathematik*, vol. 11 (1834), p. 198.

13. The first two of the pathological curves discussed here were originally constructed as examples of non-differentiable functions – continuous functions whose graphs have no tangent at any point. The snowflake curve was designed by E. Kasner in 1901, and appears in his *Mathematics and the Imagination*, New York (Simon and Schuster), 1940. An exhaustive historical development and bibliography of such functions is to be found in A. N. Singh, *The Theory and Construction of Non-Differentiable Functions*, Lucknow (Kishore), 1935.

14. W. Sierpiński, *Bulletin de l'Académie des Sciences de Cracovie*, A (1912), pp. 463–78.

15. W. Sierpiński, *Comptes rendus de l'Académie des Sciences à Paris*, vol. 160 (1915), p. 302.

16. L. E. J. Brouwer, *Mathematische Annalen*, vol. 68 (1909), p. 427. Our construction is an adaptation, due to H. Hahn, of Brouwer's original construction.

17. Galileo Galilei, *op. cit.*, pp. 31–3.

18. Proof of the fact that the number of rational numbers is A_1, while the number of real numbers is greater than A_1, is included in Cantor's first contribution to the theory of aggregates. See *Journal für Mathematik*, vol. 77 (1874), pp. 258–62.

19. It should be pointed out that the proof concerning the unit square and the unit line presents certain difficulties which were omitted for the sake of brevity. For example, our conclusion that 'there are no more points in the unit square than in the unit line' is true, but we did not show that the number of points in the square is *equal* to the number of points in the line. In other words, we merely showed that to every point P of the square there corresponds a unique point Q of the line. Certain modifications must be made in the representation of z if the converse is to be established. These difficulties are discussed in, for example, F. Klein, *Elementary Mathematics from an Advanced Standpoint*, New York (Macmillan), 1932, pp. 257–9. This book is a translation of the third German edition.

20. Proof of these results was first given by Cantor in *Journal für Mathematik*, vol. 84 (1878), pp. 242–58.

21. That the number of transfinite numbers is infinite was first established by Cantor in *Mathematische Annalen*, vol. 21 (1883). Later he gave simpler proofs of this result and of some other previous results in *Jahresberichte der Deutschen Mathematiker-Vereinigung*, vol. 1 (1890–1891), pp. 75–8.

CHAPTER 8

1. It is unfortunate that the first letter from Pascal to Fermat has been lost. A number of the later letters which passed between these two men can be found, translated into English, in D. E. Smith, *A Source Book of Mathematics*, New York (McGraw-Hill), 1929, pp. 546–65.

2. I. Todhunter gives an account of this in his *History of the Theory of Probability*, London (Macmillan), 1865, pp. 258, 259.

3. F. Galton, *Nature*, vol. 49 (1894), pp. 365, 366.

4. J. Bertrand, *Calcul des probabilités*, Paris (Gauthier Villars), 1889, pp. 3, 4.

5. J. Bertrand, *op. cit.*, pp. 2, 3.

6. The problem can be solved by the use of Bayes's theorem. See, for

example, T. C. Fry, *Probability and its Engineering Uses*, New York (Van Nostrand), 1928, pp. 121, 122.

7. Both examples are from W. Lietzmann, *Trugschlüsse*, Leipzig (Teubner), 1923 (3rd ed.), p. 16.

8. J. Bertrand, *op. cit.*, p. 4.

9. Paradox 1 is from J. von Kries, *Die Principien der Wahrschein-lichkeitsrechnung*, Freiburg, 1886. Paradoxes 2, 3, and 4 are from Bertrand, *op. cit.*, pp. 4–7. For further discussion of problems of this sort see E. Czuber, *Wahrscheinlichkeitsrechnung*, Leipzig (Teubner), 1938 (5th ed.), pp. 80–118.

10. This conclusion can be deduced from the following three theorems of plane geometry. (1) In any triangle the centre of the circumscribed circle is the point of intersection of the perpendicular bisectors of the sides. (2) In an equilateral triangle the perpendicular bisector of any side coincides with the median to that side. (3) In any triangle the medians intersect in a point which is two thirds the distance from any vertex to the mid-point of the opposite side.

11. The principle of insufficient reason is discussed at length in J. M. Keynes, *A Treatise on Probability*, London (Macmillan), 1921, pp. 41–64.

12. J. Bertrand, *op. cit.*, pp. 31, 32.

13. See, for example, the discussion by R. E. Moritz, *American Mathematical Monthly*, vol. 30 (1923), pp. 14–18, 58–65.

14. This is the St Petersburg paradox in disguise.

15. Lewis Carroll (C. L. Dodgson), *Pillow Problems*, London (Macmillan), 1894, p. 18.

CHAPTER 9

1. B. Russell, *Introduction to Mathematical Philosophy*, New York (Macmillan), 1920 (2nd ed.), p. 194.

2. B. Russell, *Revue de métaphysique et de morale*, vol. 14 (1906), pp. 627–50. See also, by the same author, *American Journal of Mathematics*, vol. 30 (1908), pp. 222–62.

3. More detailed discussions of the logical paradoxes can be found in a number of places. See, for example, B. Russell and A. N. Whitehead, *Principia Mathematica*, Cambridge (Univ. Press), 1935 (2nd ed.), vol. I, p. 60 ff.; C. I. Lewis and C. H. Langford, *Symbolic Logic*, New-York (Century), 1932, pp. 438–85; and so on.

4. C. Burali-Forti, *Rendiconti del circolo matematico di Palermo*, vol. 11 (1897), pp. 154–64.

5. J. Richard, *Revue générale des sciences*, vol. 16 (1905), p. 541. A less technical discussion of this paradox can be found in an article by A. Church, *American Mathematical Monthly*, vol. 41 (1934), pp. 356–61.

NOTES AND REFERENCES

6. An excellent discussion of trends in mathematics from the very beginning of the subject is to be found in E. T. Bell, *The Development of Mathematics*, New York (McGraw-Hill), 1940. The first part (pp. 511–536) of the last chapter of this book is devoted to the most recent investigations into the foundations of mathematics. See also T. Dantzig, *Number, the Language of Science*, New York (Macmillan), 1930, pp. 224–48.

CHAPTER 10

1. This argument has been attributed to Proclus (fifth century A.D.).

2. *Mathesis*, vol. 23 (1903), p. 133.

3. W. Lietzmann, *Trugschlüsse*, Leipzig (Teubner), 1923 (3rd ed.), pp. 44, 45.

4. *Mathesis*, vol. 10 (1890), pp. 222–4. Article signed 'P. M.'

5. W. W. R. Ball, *Mathematical Recreations and Essays*, London (Macmillan), 1931 (10th ed.), p. 51. Attributed to R. Chartres.

6. W. Lietzmann, *op. cit.*, p. 37.

7. J. L. Coolidge, *American Mathematical Monthly*, vol. 38 (1931), pp. 222, 223. The solution given in the Appendix was suggested by G. Bareis, same periodical, vol. 39 (1932), p. 29.

8. W. Lietzmann, *op. cit.*, pp. 45, 46.

9. W. Lietzmann, *op. cit.*, p. 42.

10. W. Lietzmann, *op. cit.*, pp. 47, 48.

11. W. W. R. Ball, *op. cit.*, p. 52.

12. W. Lietzmann, *op. cit.*, p. 49.

13. W. Lietzmann, *op. cit.*, pp. 50, 51.

14. W. Lietzmann, *op. cit.*, pp. 11, 12.

15. Paradoxes 2 and 3 are both from W. W. R. Ball, *op. cit.*, p. 29. The second is attributed to Johannes Bernoulli.

16. J. L. Coolidge, *American Mathematical Monthly*, vol. 21 (1914), p. 184. The solution given in the Appendix was suggested by G. Loria, same periodical, same volume, p. 327.

ADDITIONAL NOTES TO CHAPTER 3

2 (contd). In 1925 five still larger prime numbers of the form $2^n - 1$ were discovered by R. M. Robinson, using the SWAC (The National Bureau of Standard's Western Automatic Computer). They are $2^{521} - 1$, $2^{607} - 1$, $2^{1279} - 1$, $2^{2203} - 1$, and $2^{2281} - 1$. The SWAC tested the last of these numbers in about an hour, roughly the equivalent of more than 60 years of work for a person using a desk calculator.

2a. In accordance with the preceding note, five additional perfect numbers, corresponding to $n = 521, 607, 1279, 2203$, and 2281, are now known. The largest, $2^{2280} (2^{2281} - 1)$ is a number of 1372 digits.

Index